It's All on the Label

Zenas Block

IT'S ALL ON THE LABEL

Understanding Food, Additives, and Nutrition

LITTLE, BROWN AND COMPANY
Boston · Toronto

To my wife, Lil, our children, and theirs

First Edition

Library of Congress Cataloging in Publication Data

Block, Zenas.
It's all on the label.

Bibliography: p.
Includes index.
1. Food additives. 2. Food—Labeling. I. Title.
TX553.A3B58 641.3'1 81-15625
ISBN 0-316-09971-6 (pbk.) AACR2

BP

Designed by Janis Capone

Published simultaneously in Canada
by Little, Brown & Company (Canada) Limited

PRINTED IN THE UNITED STATES OF AMERICA

ACKNOWLEDGMENTS

I wish to express my gratitude to my friends Elliot Caplin, Howard Goodkind, and Joelle Sander for their support, comment, and criticism; to George Ernst, owner of Shagroy Supermarket, for his saintly forbearance over two years while I took hundreds of photos of labels; to Richard McDonough, editor, and Peggy Freudenthal, copyeditor, for their guidance, precision, and extraordinary sense of responsibility to the ultimate reader; to Cynthia Howe, who not only typed the manuscript, but demanded that I make sense; and to the Food and Drug Administration for their cooperation in supplying any material requested swiftly and efficiently.

I also am grateful to the consumer advocacy groups, food-industry scientists, FDA, and USDA for the continuous generation of information that is making our food supply safer and more nutritious — and this book possible.

Writing this book required me to read thousands of labels and volumes of legal and technical publications; to consult by phone with FDA experts; and to draw on my own experience as a food scientist and businessman. If, in the effort to simplify and present information in a manner usable by the average person, I have made any errors, they are entirely my responsibility.

PREFACE

Never before has there been so much fear, concern, interest, and confusion about the foods we eat, especially processed foods. New food-labeling regulations have resulted in food labels carrying more information than the average consumer is able to comprehend, often giving rise to anxiety and concern.

Some people believe that consumers are only confused by so much information and cannot possibly hope to use it constructively.

Others believe that consumers have a right to know exactly what goes into their food.

There is validity in both views.

In my opinion, consumers may be confused at the outset, but in time will learn to understand what processors are using in their products and why, and will be able to make their own sensible judgments about what to buy.

Consumers' advocates and the government are committed to supplying more such information and will be doing what they can to educate consumers.

The purpose of this book is to help you understand and use the information on the labels of foods — as to both ingredients and nutritional value — so that you can make informed choices about brands and, even more important, about food to help you meet your nutritional needs.

The first few chapters will provide basic background on the food laws, food safety, additives, methods of preserving foods, and the nutritional additives.

Then we will look at each food category, as you would find it in your food store, translate labels into plain English, and, where necessary, tell you how the foods are produced.

Finally, we will discuss the new dietary guidelines. In 1980, "Dietary Guidelines for Americans" appeared, developed jointly by the U.S. Department of Agriculture and the Department of Health, Education and Welfare. These guidelines are based on years of study by and discussion among the leading nutritional scientists and physicians of our time.

You can use this book in three ways:

1. To make informed buying decisions, based on your new knowledge of ingredients;
2. To plan menus and your diet so as to meet the new dietary guidelines, or to follow any diet you wish;
3. To increase your general knowledge of foods and to be more comfortable about what you eat.

There is a great deal of detailed information in this book, and it would literally be impossible for you to go shopping, book in hand, and read every label on every food you're considering buying on each trip. Read chapter 17 to learn how to make food selection simpler.

CONTENTS

It's All on the Label

Chapter 1

FOOD LAWS, ADDITIVES, AND SAFETY

□ FOOD LAWS □

The U.S. Food, Drug and Cosmetic Act, and the Meat and Poultry Inspection acts, with all of their amendments and regulations, encompass hundreds of pages, but we're going to try to boil them down to a few paragraphs.

The present Food, Drug and Cosmetic Act was passed in 1938, and has been continually amended since that time. The most important amendments are the Food Additives Amendment, passed in 1958, and the Color Additives Amendment of 1960. These amendments are intended to provide the consumer with safer foods and drugs, and more exact information about the composition of processed foods.

The basic regulations regarding labeling now are:

1. The name of the food product must not be misleading: for example, a product cannot be called "beef frankfurter" unless the meat it contains is all beef.
2. If a product has had government standards established for it — such as white bread or jam or mayonnaise — the product must meet those standards or it may not be called by that name. For example, if a preserve contains less than 60 percent fruit, it cannot be called jam or jelly. It can be called imitation jam.

3. The quantity of product must be clearly shown, by weight if a solid, by weight or volume if a liquid.
4. All the ingredients must be listed in the order of quantity present.
5. The name of the manufacturer or distributor must be shown. If not identified as the distributor, the company shown is the manufacturer.
6. If the product is artificially colored, the label must say so — with some exceptions.
7. If any nutritional claims are made, the label must provide exact information about the product's composition, stating how the product meets such claims. For example, if the label says "rich in vitamins," the law requires the label to state the percentage provided of the recommended daily allowance (RDA) of each vitamin in a serving.
8. The label must carry any required warnings. For example, if the product contains saccharin, a warning is required that says, "Use of this product may be hazardous to your health. This product contains saccharin, which has been found to cause cancer in laboratory animals."
9. Currently voluntary, but possibly mandatory in the future, the nutritional value of the food is stated on many products.

EXAMPLE: (for a canned fruit)

Nutrition Information per Serving

Serving size 4 slices with juice

Calories	140	Carbohydrate	35 grams
Protein	1 gram	Fat	1 gram

Percentage of U.S. Recommended Daily Allowances (RDA)

Protein	*	Riboflavin	2
Vitamin A	2	Niacin	2
Vitamin C	10	Calcium	2
Thiamine	10	Iron	4

*Contains less than 2 percent of the U.S. RDA of this nutrient.

ANATOMY OF A LABEL

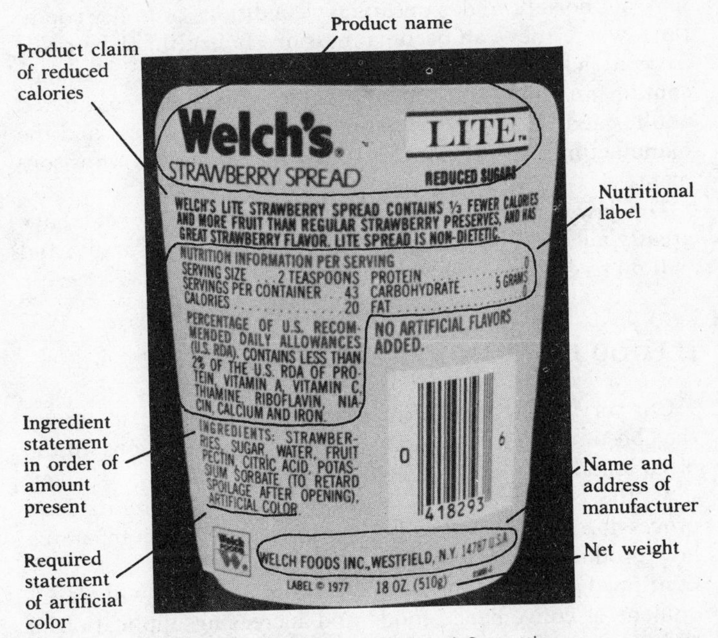

Label courtesy of Welch Foods, Inc., A Cooperative.

The label shown is an excellent example of an informative food product label. Much of the information provided is voluntary, such as the nutritional analysis. It meets the requirements for the reduced calories claim by containing one-third fewer calories than regular strawberry preserves.

The ingredients are listed by their common or usual names, in the order of amount present. The purpose of some of the ingredients is shown voluntarily.

The name and address of the manufacturer is shown. We know it is the manufacturer, because it does not state "distributor."

The net weight is shown in both ounces and grams.

The product identification is clear.

Quite apart from the labeling regulations, the law also states that no food may contain any "poisonous or deleterious substance, or filthy, putrid, or decomposed substance." Foods may not be kept under unsanitary conditions or under conditions where they can become injurious to health; or be made from a sick animal; or be placed in a container that may contain an injurious substance. Such foods are considered adulterated. They are subject to government seizure, and the manufacturer or distributor is subject to fine and imprisonment.

The law and its amendments are far-reaching and have greatly affected the purity and safety of our food supply, and will do so even more in the future.

■ FOOD LAWS AND THE FUTURE

Our very first federal food act was passed in 1906, to protect the consumer from the use of adulterants and poisons all too often found in foods.

As the country became more industrialized in its food processing and production, the need for a more encompassing law produced the Food, Drug and Cosmetic Act of 1938. Continued growth of the food-processing industry, with development of convenience foods and increasing application of technology to food processing, led to the passage of the Food Additives Amendment of 1958. (This amendment included the famous Delaney Clause, which prohibits the use of any material that is known to cause cancer in man or beast, regardless of the amount needed to cause it.) Each year, regulations under the law grow and are modified to reflect changing needs and knowledge.

In December 1979, a major step was taken by the U.S. Department of Agriculture and the Department of Health, Education and Welfare (of which FDA is part). USDA–HEW issued a statement of policy and intentions, based on extensive surveys among consumers, to make the law more responsive to consumer concerns.

Among the more important changes coming are:

1. Many foods that now do not show that they are artificially colored will be required to do so: butter, ice cream, cheese.
2. Flavors and colors that may cause allergic reactions will be specifically named on labels.
3. The specific quantity of some ingredients will have to be stated on the label, particularly of sugar.
4. Where any nutritional claim is made for a product, full nutritional labeling would be required.
5. Nutritional labeling may become mandatory, not voluntary.
6. Sodium and potassium labeling will be required. This is important to anyone whose sodium and potassium must be restricted, including high-blood-pressure sufferers.

For a complete summary, see appendix 2.

■ RESTAURANT FOODS

Despite the fact that we now spend a third of our food money in restaurants, and soon will be spending half, at present there is no requirement for restaurants to inform their patrons of the ingredients they use. In the absence of consumer pressure to minimize the use of nonfood materials and additives, many restaurants continue to use foods whose labels read like a chemical dictionary. Consumers have no way of knowing what they are being served. The FDA, in its Proposed Rules of 1979, decided not to require restaurants to declare the ingredients in their foods at this time. The reason for the FDA decision is that such a rule would be too difficult and expensive to enforce.

The question that arises from the FDA decision is this: If the needs of the consumer are sufficiently important to require the control, policing, and constantly improved labeling of packaged foods, are they to be totally ignored with respect to restaurant foods?

I think more information should be available about restaurant food ingredients. Just as many food processors voluntarily comply with recommendations of the FDA concerning packaged foods (for example, breakfast-food manufacturers *now* declare the percentage of sugar without being required to do so), many restaurants may be encouraged to comply voluntarily with sensible recommendations.

One possibility is to require restaurants to make copies of labels available to patrons upon request. For example, the patron might ask to see a label for the chocolate mix used in making hot chocolate; or the mayonnaise; or the pizza sauce that is bought by the restaurant. Product manufacturers could supply the restaurant with copies of labels or statements of ingredients used, to make it easier for the restaurant operator to supply the information on request. Failure to meet the request might result in FDA follow-up action upon complaint by the patron.

Processed foods served in restaurants are covered by the same regulations as foods packaged for consumers. What is needed now is a procedure requiring this information to be made available to the restaurant patrons.

□ WHY DO WE USE ADDITIVES? □

1. *City living:* We live in cities and suburbs, not on farms. We shop infrequently, sometimes only once a week. Since food is produced far from the consumption point and must be kept fresh and wholesome until it reaches the consumer, added preservatives are often required.
2. *Modern lifestyle:* The need and desire for refined foods and lower caloric intake require some foods to be fortified with nutritional additives to assure adequate vitamin and mineral intake.
3. *New knowledge of the relationship of food to disease* — most significantly, of saturated fats and cholesterol to heart disease — has created a need and demand for new man-made foods, which require additives to make them acceptable.
4. *High-speed processing of foods* often requires additives to make the processing economical, or even possible.
5. *More women than ever are working,* creating an enormous demand for convenience foods: prepared, ready-to-eat, or

heat-and-eat foods. Additives preserve the flavor, texture, appearance, and safety of these products.

6. *Snacking* has become a national pastime. Many snacks are man-made — with additives required to make them.

■ WHAT ADDITIVES DO

A food additive has been defined by the Food Protection Committee as "a substance or mixture of substances, other than a basic foodstuff, which is present in food as a result of any aspect of production, processing, storage or packaging. This term does not include chance contaminants."*

Substances are added to foods in order to accomplish one or more of the following things:

1. To preserve the product; that is, to prevent its deterioration from any cause.
2. To improve the texture of the food.
3. To improve the flavor, taste, or appearance of the food.
4. To improve the nutritional quality of the food.
5. To minimize loss of quality during processing itself.
6. To protect the food during its growth, harvest, and storage. These are the incidental, rather than the deliberate, materials, such as pesticides.

□ HOW ADDITIVES WORK □

Let's take a closer look at the different types of food additives and how they work.

Preservatives function to slow down or prevent the growth of bacteria, yeasts, or molds. These microorganisms may merely spoil the flavor and texture of the food or may actually produce an end product that is dangerous for human con-

*Food Protection Committee, *Principles and Procedures for Evaluating the Safety of Food Additives,* National Academy of Science/National Research Council pub. no. 750 (Washington: U.S. Government Printing Office, 1959).

sumption. Some of the more common preservatives used are sodium benzoate, sorbic acid (or potassium sorbate), and sodium nitrate.

Antioxidants slow down or prevent the reaction of components of a food with the oxygen in the air. Such reaction can produce undesirable flavors, such as rancidity in fats; unpleasant colors; and loss of vitamin value.

Emulsifiers are used for smooth blending of liquids or batters. Mono- and diglycerides are commonly used emulsifiers.

Stabilizers are often added to obtain a certain texture or to preserve a food's texture or its physical condition. For example, stabilizers are used to keep a liquid thick, to slow down the melting of ice cream, to prevent the fluid in a cheese from running off like water. Algin, xanthan gum, and other gums are stabilizers.

Sequestrants combine with trace amounts of metals that may be present in a product and prevent those metals from reacting with the foods to produce undesirable flavors or physical changes (to sequester means to keep in isolation). EDTA (ethylenediaminetetraacetic acid, a synthetically produced chemical) is a sequestrant.

Acids, alkalies, and buffers regulate the acidity of a food. We all know the difference between a tart cooking apple and a less tart (but still slightly acid) eating apple. In addition to its effect on taste, however, acidity is also very important to the preservation of food. Harmful bacteria usually do not grow in foods that are acid enough. Vitamins, including the natural vitamins present in food, tend to resist destruction more in foods that are acid, especially during the cooking process. In many foods, the preservation of the ideal flavor and color is helped by maintaining a specific acidity in the food. For these purposes, the acids make foods more acid, the alkalies make them less acid, and the buffers prevent change in acidity during storage. Citric acid is a typical acid, sodium citrate is a typical and common buffer, and sodium bicarbonate is a typical alkali.

Nutritional additives are the vitamins, minerals, and amino acids added to food either to enhance its nutritional value or to replace nutrients that might have been removed during processing. Some confusion has arisen now that vitamins are referred to by their chemical names rather than as *vitamins* in the ingredient list. It's unfortunate, but the listings of vitamins on the food label sometimes sound horrifying because their precise chemical names are given. "Vitamin B_1" is a lot more reassuring than "thiamine mononitrate."

Colors and flavors are added to make the food more appealing in appearance, smell, and taste.

Bleaching or maturing agents serve to oxidize wheat flour. While many foods must be protected against oxidation in order to preserve their quality, wheat flour used for baking must be oxidized in order to achieve the necessary quality.

Many years ago, it was possible to store flour for the necessary period of a month or longer to allow such oxidation to occur naturally with air. However, with bulk handling of flour and the massive bakeries that exist today, much flour is bleached in order to achieve artificial aging and to make the flour suitable for industrial use so that bread and cake can be produced uniformly at high speed. (I'll discuss bleaching in more detail in chapter 7.)

Non-nutritive sweeteners sweeten without calories. Until 1978, both calcium cyclamate and saccharin were used to replace sugar in foods for those people who were supposed to restrict their intake of ordinary sweets. When cyclamate was "found" to produce cancer in laboratory animals, it was prohibited under the Delaney Clause. Since 1978, saccharin alone has been used.

For years, saccharin was required to bear the statement "Contains *x* percent saccharin, an artificial sweetener to be used only by persons who must restrict their intake of ordinary sweets." Sometimes this statement was in type so small it could not be read without a magnifying glass.

Now, saccharin also has been "found" to produce cancer in laboratory animals. Under the Delaney Clause, it too would have been prohibited, but since it is the only usable artificial sweetener available, a delay in its prohibition has been granted, and even extended, to provide more time for confirmatory testing. (The granting of this delay actually required Congressional action.) In the meantime, saccharin must carry the warning: "Use of this product may be hazardous to your health. This product contains saccharin, which has been determined to cause cancer in laboratory animals."

The history of artificial sweeteners is an example of the seriousness with which our food labeling regulations are applied, and of the problems raised by the Delaney Clause. The rationale for the delay in the prohibition of saccharin is that the danger to people that would occur through its elimination (overweight and resultant cardiac disease and stroke) is greater than the danger of cancer. In the meantime, retesting of saccharin and cyclamates, and the testing of new substances, such as aspartame, is proceeding in the hope of finding that there is no confirmable cancer-producing hazard.

Miscellaneous materials such as leavening agents and those used for other special purposes will be explained as we go through the labels further on.

■ REDUCING COSTS WITH ADDITIVES

The food business is tough and competitive, and profit margins are low compared to other industries. Most food companies earn a profit of less than five cents for every dollar of food sold. In addition, because competition is so keen, sometimes a price differential of a penny or two for a food product is sufficient to make a difference in the market share a product will achieve. These pressures prompt food manufacturers to find ways to reduce their production costs, sometimes by using new methods of processing, as well as new formulas or recipes for the food. Very often a recipe is developed that enables the cost of a food product to be reduced but that requires the use of additives in order to achieve the quality of the original product. Sometimes a food product that is totally natural and simple, such as maple syrup, becomes so costly to produce that its price is discouraging. A cheaper maple-flavored syrup is made from a sugar-syrup base. Stabilizers are added to make it thick, artificial flavor is added to make it taste roughly like maple, artificial color is added to make it look approximately like maple syrup, acid and buffers make the acidity something like that of maple syrup.

Often additives are necessary because the cost of a food has been reduced by diluting it with water, after which the additives are used to restore consistency. Examples of this practice will be pointed out as we read the labels.

□ WHAT ABOUT SAFETY? □

Now that we have a smattering of knowledge about the laws and the world of food additives, we can begin to address the most important questions: Is our food safe, and how do we know it? Are so many additives necessary? Are we taking unnecessary risks? Somehow we feel differently about the safety of what we eat than we do about the safety of other elements in our lives. Are airplanes safe? Certainly not completely, but safe enough for millions of people a year to make up their minds to get into an airplane and fly. Are automobiles safe? Far from 100 percent safe. More people have died in automobile accidents in the United States than in all the wars the United States has ever fought. Yet millions of us get into automobiles every day and drive them.

But there is a difference between choosing to get into an automobile or an airplane and having additives in foods. We make the individual choice, more or less, of flying or driving. We understand that there is a risk. We can balance our understanding of the risk against the benefits we gain, and we make the decision. But that isn't the case with food. We really don't understand the risk in all those "chemicals" in food. We may try to choose "all-natural" foods, but their prices and tastes frequently do not meet our needs, and there is *no* evidence that they are less risky.

Moreover, when one gets into an automobile and takes a trip and gets out of the automobile, the danger or the risk of that trip has ended. We are finished. There is no lingering effect.

That is not so when we eat a food. If we eat a food containing a material that might cause cancer, it doesn't do so on the day we eat the food.

There is a great deal of worry about what our foods consist of, usually quite unjustifiable. It is certainly time for us to learn more about the risks versus benefits in our food supply, and to make individual judgments that are based on knowledge rather than blind faith in either the food processors or the natural-food advocates.

We can start with a very simple statement: *too much of anything is toxic, that is, poisonous. The only thing that varies is the dose required to produce the toxic reaction.* This in turn must mean that nothing is absolutely safe. Furthermore, people vary greatly in their tolerance for substances. An obvious example is the case of common table salt: high-blood-pressure sufferers have a toxic reaction to much less salt than people with normal blood pressure. Other examples are seen in allergic reactions by some people.

Unfortunately, most foods have not been safety tested. While continued use over many generations indicates that there probably is no acute danger, it does not *prove* that the food is nontoxic or safe to an acceptable degree. Charcoal-broiled meat has recently been found to contain cancer-producing substances; nitrates have been used for more than a thousand years, yet only in the past fifteen years have they been found to produce carcinogens.

Furthermore, there is no conclusive way to test food safety, particularly since the tests must be conducted on animals. The Code of Federal Regulations, Number 21 (Food and Drugs), states, "To provide assurance that any substance is absolutely safe for human or animal consumption is impossible. This is particularly true for substances intended for human consumption which have been tested in animals." However, the object of food-safety tests is to determine, as the Code of Regulations further states, that "no significant risk of harm will result when the substance is used as intended." The safety judgment is made after reviewing all available evidence, including how much of the substance will be consumed, cumulative effect in the diet, and safety factors determined by trained experts.

The Food Safety Council, governed by a board of trustees composed of equal representation from industry and the public sector, was formed in 1976 to develop criteria for determining the safety of food substances. The council has proposed that a substance be regarded as safe "if it presents a socially acceptable risk of an unfavorable effect at levels of consumption that are experienced by high consumers of foods in which the substance occurs."*

This suggested definition is part of a proposed system for food-safety assessment that, in a highly systematic and thorough fashion, emphasizes objective scientific information covering effect on heredity, whether the substance is chronically toxic, the way the substance is metabolized in the body, methods for judging how much of the substance is likely to be used, and methods for calculating risk.

The "socially acceptable" level is not to be determined by scientists, but by the Food and Drug Administration (part of the Department of Health, Education and Welfare).

Currently, substances that may be added to foods are permitted up to a level no greater than one one-hundredth of the level found to be safe in animals, unless convincing evidence is found to justify a change. The term *level* refers to the amount of the substance that will be ingested per pound of body weight.

Mankind has been eating foods without knowing what they contain for millions of years. As our knowledge of the composition of foods increases, it is possible that we will change our choices of the foods we eat. This process has already begun and will become more intense in the years ahead.

What appears to be much more significant than the substances added to our foods is the composition of our diets, that is, the amount of protein, fat, carbohydrate, fiber, vitamins and minerals, and calories that we eat. While we have been developing more and more processed foods, with more and more additives, our life spans and years of vigor have increased, either because of or in spite of the changing nature of

*Food Safety Council, *Proposed System for Food Safety Assessment: A Comprehensive Report on the Issues of Food Ingredient Testing,* new ed. (Elmsford, New York: Pergamon Press, 1978).

our food supply (or for reasons having nothing to do with our diets). In the decade ending in 1980, the death rate owing to heart disease declined by 27 percent. It is believed that this is a result of our reduced consumption of saturated fats and cholesterol, and greater awareness of and attention to other factors, such as smoking and exercise.

■ WHAT ABOUT THIS "CHEMICALS" BUSINESS?

Some people are concerned about the chemicals in our foods, and in the past few years, there has been a tremendous boom in so-called "natural" or "organic" foods, because they are free of "added chemicals."

The fact is, any substance is nothing more nor less than a combination of chemicals. The use of the word "chemical" to describe a substance is meaningless, since everything is chemical.

If milk required an ingredient statement, it would read like a chemical-factory inventory: "Sodium caseinate, calcium caseinate, lactic acid, triglyceride esters, retinol, sodium phosphate," and so on. This would be equally true for lettuce, soybeans, or any other naturally occurring food.

The statement "Without chemicals, life itself would be impossible," a slogan used by a chemical company to help take the curse off the word "chemical," is true enough. We should realize that, very simply, there *isn't anything* that isn't a chemical, except energy itself (heat, light, electricity).

Nevertheless, there are materials in our food supply, introduced by man, that have frightening chemical names. In spite of all the safety testing done, there is no such thing as absolute safety, and the safety tests that have been done on many substances were performed when testing techniques were less well-developed than today's. It will take some time to test more of them using new techniques, a process that is going on constantly.

What are we to do in the meantime? Should we believe that everything permitted by the government is safe enough? Before suggesting an answer to this question, it is useful to look at chemicals in foods in another way. The chemicals we eat, whether man-made or naturally occurring, fit into one of the following categories:

1. They may be a normal part of our bodies. Salt is such a chemical. Such chemicals, if consumed in quantities that are a small percentage of the amount in the body and by persons who are not required to restrict their intake of such materials, are the most truly natural chemicals of all. It is difficult to imagine anything safer.
2. They may be converted by digestive processes into materials that are a normal part of the human body. An example is sugar, which is converted to glucose through digestion. Glucose is part of the human body. However, as we know, diabetics have a problem with the absorption and handling of glucose. There is also evidence that too much sugar for normal people may have poor effects on health, for reasons not yet explained.
3. The chemical may be converted into a substance that is harmless, although not a normal body substance. Sodium benzoate is an example of such a chemical. (Note: sodium benzoate is commonly found in many berries, i.e., it is "natural.")
4. The chemical may be converted after digestion into a normal body substance by the metabolic processes in the body.
5. The chemical may be consumed and accumulated in quantities too small to produce a toxic reaction.
6. Substances may be completely excreted without change of any kind. (Saccharin was believed to be a substance of this type.)

Now we come to another problem. How can we determine what really happens to the chemical once it enters our bodies? We need to be able to analyze blood, urine, feces, and tissue to check for the presence of the chemical, as well as any substances it may have been converted to. In 1958, when the Food Additives Amendment was passed, it was generally possible to detect substances present in amounts of a few parts per million (one pound of salt in one million pounds of water is one part per million). With new instrumentation, it is now possible to detect substances in amounts of a few parts per *billion*—a thousand times more sensitive!

Instead of months and years of tests on animals alone, recombinant DNA studies can produce the effects of a substance on genes in weeks, even days, by using bacteria as a test medium. In this way, materials that cause genetic changes can be quickly screened out.

This new testing power enables us to find substances in smaller quantities than ever before, faster than ever before, and to determine toxicity more thoroughly and more rapidly than ever before. It is being used to a maximum extent to learn as much as possible, as fast as possible, about the elements of our food supply that demand immediate attention.

To summarize: *There are no substances in use that are known to be unsafe under the required conditions of use,* but there are important differences in how sure we are about their safety. Until we know more, the prudent person will avoid excessive amounts of any one substance, and select foods, when there is a choice, that have fewer additives—especially avoiding those substances that are not normally found in the human body or converted to such. This book will assist you in identifying many of the normal body chemicals sometimes used as food additives.

□ THE CLASSIFICATION OF FOOD SUBSTANCES AND ADDITIVES □

The Food Additives Amendment was passed in 1958 to bring some control and order to the relative chaos of materials being added to foods. Here is how this was done. Three major classifications were made:

1. Substances Generally Regarded as Safe (GRAS)
2. Additives
3. Substances that are prohibited

Substances that were Generally Regarded as Safe consisted of all of the foods of natural, biological origin that had been in use prior to 1958, plus substances that were considered safe by experts familiar with the scientific work done with those substances, their history, and their use.

Substances were considered additives if there was not a general opinion that enough was known about them to consider them safe. Evidence of safety was required, as described earlier, but the experts did not have the same degree of certainty regarding safety as for the GRAS materials. Included in the additives group were some substances approved before 1958.

Prohibited substances were those eliminated from the food supply because they were found to be toxic. Two examples are coumarin in vanilla flavor and, more recently, calcium cyclamate as an artificial sweetener.

It was recognized in 1958 that the status of substances classified as GRAS could change, but it was not until 1969 that a systematic reevaluation was begun to either affirm a substance as GRAS or determine whether reclassification as a food additive was required.

Priorities were set as to which substances would be examined. The plans for retesting were impressive and reassuring. For example, any natural substance used as food would be reexamined if its processing had been changed since 1958 in a way that might affect its composition or nutritive value. Any food would be reevaluated if it had been altered by breeding (such as a new corn variety), if the change might be expected to alter nutritive value or concentration of toxic substances. Components of GRAS substances were identified for future checking. These are natural substances used in food, but not for their nutritive properties. An example is vegetable color.

In April of 1980, the Federation of American Societies for Experimental Biology submitted a report on the results of massive studies conducted over a period of eight years on 415 substances, most of which were GRAS, but a few were additives, such as modified food starches. This work was done under contract for the FDA. The conclusions were reached by a select committee that consisted of thirteen experts: eminent professors at a number of universities in the fields of pharmacology, chemistry, pathology, pediatrics, nutritional biochemistry, community medicine, nutrition, and oncology (the study of cancer).

The Select Committee on GRAS Substances (SCOGS) based its evaluation of safety on a review of medical and scientific publications as well as on unpublished reports. The number of reports for each substance varied from as few as 23 for one to as many as 20,000 for another.

SCOGS developed five classifications, into which each substance studied could be placed. Those classes are:

> *Class 1*
> Continue in GRAS status with no limitations other than good manufacturing practice.
>
> *Class 2*
> Continue in GRAS status with limitations on amounts that can be added to foods.
>
> *Class 3*
> While maintaining GRAS status, tests must be conducted within a specified period. GRAS status remains until tests are conducted and evaluated.
>
> *Class 4*
> Establish safer usage conditions as a food additive (end GRAS status), or remove the ingredient from food.
>
> *Class 5*
> The data are insufficient to make an evaluation. We can't tell whether it is safe enough to use as a food ingredient or not.

These conclusions are based on the quality and quantity of evidence regarding safety of the ingredient. SCOGS was concerned not only with present usage levels, but possible future usage levels as well. Some of the statements are complex and difficult for the nonscientist to follow; for example, the basis of class 4 is: "The evidence on substance or substances is insufficient to determine that the adverse effects reported are not deleterious to the public health when it is (they are) used at levels that are now current and in the manner now practiced."* Translation: Some bad effects are reported, but we can't tell for sure that those effects won't hurt the public, at the amounts now used, so we had better find out how much is safe, or get the substance removed from the food.

A simpler way to think about the classes is as a clue to how sure we can be about the safety of a substance. A classification of "5" doesn't mean that the substance is dangerous — but it might be. We are least sure about its safety. We are most sure about a "1," but since absolute safety cannot be proven, all we can say is that we are surest about class 1, not that substances classified as such are absolutely safe.

Finally, even though thousands of reports have been studied, the results cannot be considered the final word — merely the best to date.

It would seem logical that consumers should know whether substances are GRAS or additives, and perhaps, with this most recent information, what class of GRAS the substance fits into.

In this book I show the class for each substance mentioned. Where there is no class shown, it means the substance has not been classified yet.

You will note that there are some natural substances that are *not* in class 1. The fact that a substance is natural does not make it GRAS, unless it was widely used as a food before 1958, and regarded as safe by experts, without any evidence of harm. If you wished to introduce a new food, for example, krill, a shrimplike creature now eaten by whales by the ton, you would have to make application for additive status, with proof of safety under the proposed conditions of use, and then later apply for GRAS status.

After using this book for a while, consumers will be familiar

*From the report of the Select Committee on GRAS Substances (SCOGS), 1980. Unpublished, available from FDA.

with many of the GRAS materials and what class they are in, as well as the additives. For the time being, I recommend using products containing the fewest number of nonfoods, and, if I had a choice, would select products that contained GRAS rather than additive materials, preferably those in GRAS classes 1 and 2.

I haven't listed class 1 substances because: a) There are hundreds of them, and b) There is least reason for concern about them.

You will note that sugar is listed in both class 2 and class 4. This is because there is no doubt that sugar is related to cavities, and for that reason is listed in class 4. Aside from that, there is a need for more study about the effects of sugar on health when used in large quantities, hence the listing in class 2 as well.

Guide to the Most Commonly Used Materials in Classes 2 through 5

Class 2
Carob-bean gum
Guar gum
All the sulfites and sulfur dioxide
Agar-agar
Alginates
Methyl cellulose
Licorice
All forms of iron used as nutritional supplements except ferric oxide
MSG and other glutamates
Dextrose
Corn sugar
Invert sugar
Sugar
Vitamin D
Many modified starches
Vitamin A

Class 3
BHT, BHA

Carrageenan
Nutmeg, mace, and their essential oils (natural flavors)
Caffeine — in cola drinks
A number of modified starches

Class 4
Salt
Sugar — because of dental caries (cavities)

Class 5
Ferric oxide — as a nutritional source of iron
Manganous oxide — as a nutritional source of manganese

□ FOOD ADDITIVES AND CANCER □

We recognize that even "harmless" substances can be poisonous, given a large enough dose. But when it comes to substances that "cause" cancer, any dosage is too great, as far as the present law is concerned. The Food Additives Amendment provides, in the Delaney Clause (Sec. 409), that "no additive shall be deemed to be safe if it is found to produce cancer in man or animal." Any case of cancer, no matter how rare, no matter what the dose involved, that can be considered to be caused by the material, is sufficient to rule out the use of that material as an additive.

Under this ruling, we have seen calcium cyclamate eliminated, and saccharin temporarily permitted only by special Congressional action. Also, a number of artificial colors have been eliminated.

Such a provision is neither hysterical nor arbitrary. There is some evidence that once a single cell is changed by a cancer-producing agent, it can grow into a tumor. This justifies the point of view that dosage is not crucial. Another view is that cells can repair themselves and the body has the ability to reject potentially abnormal cells, thus it is possible to set a dose threshold.

There are two major cases of additives that have been

suspected of causing cancer: the nitrates (and nitrites) and saccharin.

■ THE NITRATES AND NITRITES

Nitrates have been used for thousands of years as meat preservatives, initially in the form of saltpeter, a mineral substance occurring in nature.

Nitrates not only preserve meat, but also provide the attractive red color that we associate with ham, bacon, and hot dogs. Bologna, sausages other than frozen sausages, and liverwurst also contain sodium or potassium nitrite.

Occasionally, products will contain both nitrate and nitrite. For many years, only nitrate was used. It was then discovered that during the curing process, the bacteria in the curing solution converted the nitrate to nitrite, and that it is the nitrite that performs the change in color and the preservation. Blends are still used by some manufacturers to regulate the rate of curing. Most commonly, however, sodium nitrite alone is used.

The cancer concern has been related to two specific factors: (1) nitrosamines, which are formed by nitrites reacting with amines (substances naturally found in meats) under certain conditions, and (2) the nitrites themselves.

It was in the early 1960s that it was discovered that nitrites could form nitrosamines and that these nitrosamines do cause cancer of the lung, digestive tract, liver, and nervous system. In 1973, it was confirmed that nitrosamines can be found in bacon after frying. At that time a panel was appointed to evaluate the problem and hazard and to recommend action. In 1978, the panel recommended that nitrites be reduced. (Since nitrite was in use before the passage of the Food Additives Amendment, it has prior sanction and the Delaney Clause cannot be applied.)

The panel could not recommend elimination of nitrites without courting disaster. Nitrite curing of nearly 7 *percent of our total food supply* prevents botulism, a deadly disease. Presently only ten to twenty cases per year occur in the United

States, mostly from home-processed canned foods. *There has never been a case of botulism caused by commercially processed meat in the United States.*

Beginning in 1977, a number of actions were taken, including the following:

1. The meat industry was asked to submit data showing that they could produce meat that contained no nitrosamines when prepared for eating.
2. In 1978, the level of nitrite that could be used for curing bacon was reduced from 200 parts per million to 120 parts per million, and sodium ascorbate or sodium erythorbate would be added to prevent formation of nitrosamines. Under this regulation, cooked bacon could not contain nitrosamines in amounts that could be found.
3. A proposal was made to reduce nitrite in bacon to 40 parts per million, and use it in combination with potassium sorbate, which produces less nitrosamine and provides protection against botulism. Testing of this continues.
4. In December of 1978, USDA began regular testing of commercially produced bacon for nitrosamines. Bacon made by injecting nitrite cures is required to be free of nitrosamines when tested by a method that can detect as little as 10 parts per billion!
5. In September of 1978, regulations were introduced permitting the sale of nitrite-free bacon, corned beef, ham, and so on, but a warning label was required: "*Not* preserved —keep refrigerated below 40 degrees Fahrenheit at all times." But in February 1980, a permanent injunction was obtained to prevent calling these products "bacon," "ham," and so on, unless they were cured in the normal manner. The argument is that people would be misled into thinking that the nitrite-free "bacon" could be handled like regular bacon, in spite of the warning, possibly resulting in poisonings and destruction of the bacon business for all producers.
6. As a result of all the testing, in June 1980, the USDA announced that most nitrite-cured products, such as sausages, corned beef, ham, and hot dogs, do not form nitrosamines.

Nonetheless, the USDA and other agencies are searching for a nitrite substitute: either other chemicals, or processing changes, or both. Industry is doing the same. In fact, as indicated, there is already salt-cured "bacon" with no nitrite now being marketed.

So, on the question of nitrosamines, the hazard has been reduced and efforts are being made to avoid the problem completely.

On the matter of nitrites themselves, a study completed in 1978 by MIT indicated that the incidence of cancer of the lymph glands in rats that were fed nitrites was twice as high as in a control group that was not. Some 2000 rats were used in the study and over 50,000 samples of tissue were examined.

As a result, FDA considered phasing out nitrite on that count alone, but another evaluation was made, in which the same tissue samples were examined, this time by trained pathologists who did not know which tissues were which. Their only task was to identify cancerous tissue in the samples. This second group found far fewer incidences of cancer. The number they found was so low that the concern was reduced to the point that no action to phase out was recommended. What was recommended was that all data on nitrite be reviewed before additional studies are considered.

Therefore, on those two counts — nitrites and nitrosamines — we now have far less cause for concern than we had previously.

This entire history certainly is reassuring to me because it shows that important issues that require attention get it. However, consumers should continue applying pressure to assure that such attention continues.

■ SACCHARIN

Back in 1977, the FDA proposed to restrict the use of saccharin in diet sodas and other diet foods. This was a result of three studies, two in the United States and one in Canada, which showed that saccharin could cause cancer in test animals at high doses. Congress temporarily prohibited the FDA from carrying out this restriction and ordered further study. The warning labeling required for saccharin was also introduced at that time.

In December 1979, findings from a study carried out by the National Cancer Institute were made public. That study showed that there was *no increased risk of bladder cancer among users of artificial sweeteners in the overall population studied.* However, there was some evidence that sweeteners might be hazardous to three special groups:

1. Heavy users, particularly those who consumed both diet beverages (16 ounces per day or more) plus six or more servings per day of sugar substitute. The risk of bladder cancer was increased 60 percent in this group.
2. Heavy smokers (more than two packs a day) who also used artificial sweeteners. The extent of risk increase was not defined.
3. Women who consumed sugar substitutes or diet beverages twice or more per day had a 60 percent greater chance of bladder cancer than women who never used artificial sweeteners. It should be mentioned that bladder cancer is one-third as common in women as in men.

How big is that 60 percent chance for the higher risk groups? About nine cases more per year per 100,000 people. If everybody in the country were in the higher risk groups, it would mean an increase of about 2200 cases per year among our 250 million people. This is very small, but no comfort to the victims.

The increased risk of bladder cancer to males, from average use of sugar substitutes, was 18 percent — not the 60 percent originally estimated from the first three studies, which led to the FDA request to ban saccharin.

At this point, let me illustrate the foolishness of reacting to percentages alone. If one person per year in our population died of a certain disease, and consumption of a food additive increased the death rate by 100 percent, that would mean only two persons per year would die of that disease — a totally negligible risk. On the other hand, if 100,000 people died of a disease, and the food additive increased the death rate by 30 percent, then 30,000 more people would die. Remember to consider the numbers as well as the percentages!

Although the saccharin problem seems to be less serious than originally thought, there still is some evidence to concern us. We don't know enough about long-term use, and moderation is recommended.

■ PERSONAL CONCLUSION

This is a good place to make an important point. There is a difference between deciding what the government ought to do and what we as individuals should do. The government *must* be concerned about our *individual* health. Because the government, as a matter of public interest, after weighing the risks and benefits, decides to permit a substance to be used (such as nitrites and saccharin), this does not mean that we as individuals should think that it is either necessary or desirable for each of us to consume indiscriminately foods containing those ingredients. If, after reviewing the information, I decide that I would rather eat nitrite-containing foods less frequently, or not at all, and replace such foods with higher priced foods, that's a personal judgment about what I ought to do, not what the government ought to require.

Similarly, if I think I can control my weight without the use of saccharin-sweetened sodas, or saccharin tablets in my coffee, that's a choice I can make. Others may not wish to or may not be able to make that choice. In those cases, the risks of overweight probably are far greater than the risks associated with saccharin usage.

My personal point of view about all nonfood materials in foods, and indeed, even the food materials, is to make sure that I don't eat too much of the same thing—to keep my diet diverse—and to avoid materials about which we know the least. I'm prepared to read labels and study information to be able to do that. This does not mean that everybody can or should do the same thing, except that diet diversity is almost always preferable to concentration on just a few foods.

□ COLORS □

Colors are treated separately under the law and are controlled by the Color Additives Amendment of 1960. See chapter 4 for a complete discussion.

□ PESTICIDES AND OTHER INCIDENTAL ADDITIVES □

Along with additives used for the purposes stated above, incidental additives also are found in our foods. These incidental additives have two main sources. First, since our population has doubled in the past forty years while the amount of land devoted to farming has shrunk, farming has had to become tremendously efficient. However, the herbicides, pesticides, and fertilizers that promote this efficiency find their way into our foods, not only in crops but in our farm animals and poultry as well. Second, the enormous industrial output of the world has created wastes going into the ground, the air, rivers, and oceans, with occasional contamination of natural food materials (an example is mercury in tuna).

No label declaration for these materials exists; their use is governed by the FDA's tolerance standards, which determine the maximum limits for the amounts permitted in specific foods. This includes packaging materials as well, even to the components of the ink used. Many materials are prohibited, and undoubtedly there will be more prohibitions in the future.

Traces of pesticides in our food is the price we must pay for living in a highly populated world. In spite of all the pesticides used to destroy insects, to kill weeds, to kill rats and mice, to prevent mold growth, we still lose 25 percent of our total food crops to pests! While it may be possible to grow a small kitchen crop "organically," we absolutely need the high yields per acre pesticides help to give us in order to feed ourselves and the world.

Even a so-called organic product may contain pesticide contaminants, such as herbicide carried to a crop by irrigating water, or a weed killer previously used on the land, or other possible sources. Since there is no label requirement to state the amount of pesticide residue in a food, there is no objective method for consumers to learn which foods they purchase contain any residues and which do not, whether they buy the product in a health-food store or elsewhere.

Quite apart from the matter of sufficient food supply, pesticides are also necessary to prevent some food crops from being contaminated by more potent natural poisons than pesticides, such as ergot, which contaminates barley or rye, or diseases carried by rats and deposited with their excreta.

All we consumers can do is make sure we wash fruits and vegetables thoroughly before consumption and wait for future developments that may result in the use of fewer and safer pesticides.

It should be some comfort that our techniques for detection are now much more sensitive and faster than ever, which improves the ability to police pesticide use and residues. Perhaps some day food labels will be able to state: "Contains less than *x* parts per billion of pesticides."

Chapter 10 will deal with incidental additives in meats.

Chapter 2

HOW FOODS ARE PRESERVED

Any food that has been harvested and that people aren't ready to eat must be stored or discarded. Most foods can't be stored without spoiling, which means they must be preserved to be stored.

Only when mankind learned how to store grains, especially wheat, did it become possible to settle in one place rather than to wander continually in search of food. Today, food is grown far from where most people are, and there is simply no way to get it to the people without some kind of preservation. Without food storage and preservation, we could not support the world's population and could not have anything resembling urban civilization. This has been true since ancient times.

But are we adding too many materials—"chemicals" —that not only may not be necessary, but may even be harmful? Are inferior foods being doctored—for the profit of a few? To make this judgment, we must understand how food spoilage occurs and how it can be prevented.

□ WHY FOODS SPOIL □

From the moment a food is harvested, it begins to change. It can change in *taste, texture, appearance, safety,* and *nutritional value.* Not all such changes are unwanted. Indeed, we deliberately age meat to improve its tenderness and taste. There are many agents that produce those changes.

Microorganisms

The most important agents of changes in foods are the microorganisms (too small to be seen with the naked eye — a microscope is needed). We are concerned with three kinds of microorganisms: bacteria (germs), yeasts, and molds. They grow in or on foods when conditions are favorable. They change taste, color, and texture, but most important, the harmful microorganisms produce poisons that can and do sometimes kill people. Botulism, usually a fatal disease, is caused by infection of food with botulinus bacteria. A certain mold produces aflotoxin, another poison. Yeasts are not harmful, but they can spoil the taste, texture, and color of food.

But we can immediately think of certain foods that are produced or preserved by friendly, even necessary, microorganisms.

Pickles are made by bacteria and preserved by the lactic acid and acetic acid produced by the bacteria.
Bread requires yeast to make it.
Wine can be considered preserved grape juice, with its sugar turned into alcohol by the yeasts.
Wine vinegar is produced if the wine does not have enough alcohol or is contaminated by vinegar-producing bacteria. The vinegar is then preserved by the acetic acid produced by the bacteria.
Cheeses are made by using the specific bacteria and molds that give each cheese its distinctive appearance, texture, and flavor.

Enzymes

All living things contain enzymes, catalysts that regulate the hundreds of chemical changes that occur during their lifetime. (A catalyst helps a change to take place, but doesn't change itself.) These enzymes are in delicate balance, but once the organism is dead, the balance is destroyed, and deterioration of the food begins: in flavor, texture, and color. The aging of meat to tenderize it is accomplished by simply permitting enzymes to work. But the rotting of tomatoes and other fruits and vegetables is also caused by enzymes.

Oxygen

The oxygen in the air reacts with components of food, usually to produce a change in flavor or color. Rancid fat is merely fat that has reacted with oxygen. Rancid potato chips are the result of the fat on the surface of the chip reacting with the oxygen in the air. Note carefully: the substances formed that make food rancid are not particularly healthy and should be avoided.

Loss of Moisture

Many fruits and vegetables contain more than 90 percent water, which is maintained when they are in their natural state. Immediately after harvest, they will begin to lose water, and wilt.

Light

Light causes color fading and vitamin loss in some foods. Ultraviolet light promotes oxidation, and therefore its effects on food are similar to those of oxygen's. That's one reason amber bottles are used for some products, although not the only reason. (Another is that some fruit juices aren't very pretty when seen in a clear glass container.)

In order for these agents to be able to produce the changes in taste, texture, color, safety, and nutrition, the *conditions* have to be right. Conditions such as temperature, humidity, water content and acidity of the food affect how much and how fast the agents will spoil the food. There are natural "poisons," or chemicals that prevent the microorganisms from growing, if indeed microorganisms are present. Some enzymes need minute traces of other chemicals in order to work, and some chemicals either speed up or slow down changes. For example, the tiniest amount of copper — as little as one part per million — in a fat can speed up spoilage owing to rancidity a thousandfold!

In the case of microorganisms, the most important condition is the size of the microorganism population present in the first place. Food prepared under dirty conditions will be loaded with them, and spoilage will be much faster than in food prepared under clean conditions.

The principles of food preservation are simple: either *keep out agents,* or *create conditions where they do little harm,* or *some of both.* There are three fundamental ways to do this:

1. Physically
2. Chemically
3. Using microorganisms themselves (This is the same as chemically, except that the chemicals are *produced* by the microorganisms.)

Each of these three methods will have an effect on all food properties. The effect will not necessarily be positive. Usually, one or more features of the food will be sacrificed or compromised in order to preserve a more important feature, or what somebody thinks is more important, or what somebody thinks *you* think is more important!

Take the example of preserved tomatoes. They can be canned, sealed, and heated to kill all the microbes. They no longer resemble the fresh tomato in appearance, texture, or flavor, but they do retain the basic nutritional value, and they are really safe. Occasionally you might run into a label that shows that calcium chloride has been added. This is done in order to firm the texture of the tomato slightly while it is being heated, so that the tomato will at least stay in one piece. Very often, as we will find when we go through labels later, additives are used to improve a feature that is otherwise hurt by the preserving process.

The method of preservation used for a food is usually chosen to meet a quality and convenience standard at a cost that the food processor believes will be accepted by the end user — that's us!

Sometimes, many methods of preservation are used to produce a variety of products because there is enough demand for each product. Orange juice is an excellent example. The most inconvenient and expensive source of orange juice is fresh oranges. Plenty of fresh oranges are sold to people willing to pay more money and take more trouble for the quality they want. Lower in price is refrigerated orange juice, shipped from Florida or California in tank cars or trucks to packers and then to your local supermarket.

Frozen concentrated orange juice is next lowest in price, and is a little less convenient than the refrigerated, and different in taste. In addition, there is canned orange juice. Not much is sold: it is a holdover from the days before frozen and refrigerated articles were available. Finally, for bakers and other food processors, there is dehydrated orange juice, for use as an ingredient in the foods they process.

In this one food we can see many of the physical forms of preservation used: freezing, refrigeration, canning, and drying. Over the next few pages we will discuss these and other physical methods in more detail.

□ PHYSICAL METHODS OF PRESERVING FOODS □

■ DRYING

This is one of the oldest methods known. Primitive peoples dry fish in the sun to preserve it. For centuries, figs, raisins, and dates have been sun dried to preserve them, as have many herbs and spices.

Drying acts to preserve by removing the water necessary for microbes to grow and for many enzymes to act. (Blanching — that is, heating briefly in boiling water before drying — will destroy enzyme action and make the dried food keep longer.) Also, the absence of moisture in the food prevents most chemical changes from taking place. However, depending on the method of dehydration, foods change color, texture, and flavor, and they lose vitamin strength and tend to become rancid sooner. The foods that we are familiar with and accept as dried foods are those that survive the process best: peas, beans, and spices.

Most foods contain between 70 and 90 percent water, and vegetables usually between 80 and 98 percent water. The transportation cost of shipping water from California to New York in string beans, carrots, and lettuce is enormous. Efforts to dry these foods have been made, but only in wartime and only for soldiers have they ever been produced for direct consumption. However, many vegetables are dehydrated and sold to other food processors for use in dry soup mixes and other foods.

The most common method used for the removal of water is *tunnel drying:* tremendous amounts of hot air are blown into a tunnel through which the food is traveling on a conveyor. The food goes in wet and comes out dry. Most vegetables and herbs are dried this way.

Liquids and purees are dried by *spraying* them in a fine spray into an enormous tower, perhaps 100 feet high and 40 feet wide. Hot air is blown into the tower. By the time the droplet reaches the bottom, it has dried, and the particles of dry material are collected. The procedure takes seconds. Instant coffee is often dried this way, as is orange juice.

Thick pastes and purees are dried on a hot rotating cylinder. By the time one revolution of the cylinder is completed, the product is dry and is scraped off the drum by a knife blade. The product comes off in sheets and may be broken up into flakes. That's how dried bananas are made, as well as instant potatoes, perhaps the dried vegetable most successful for home use.

Freeze-Drying

Freeze-drying is the most important advance made in drying-preservation technology in the past fifty years. With this method, the food is flash-frozen and then dried directly from the frozen state under high vacuum. At no time does its temperature go higher than freezing. This does minimum damage to flavor, color, and texture. The process was originally developed to preserve tissues for medical study and was later adapted to foodstuffs. Because of the high cost of the process, it can only be used economically for products that are already high-priced, such as instant coffee and shrimp. The freeze-drying process is also used to make dehydrated campers' foods.

Frying

A popular form of dehydration is frying. Potato chips are slices of potato that have been fried to dryness. This is how many other fried snacks are made. Many of them contain BHT and/or BHA, materials that are added to fat to prevent rancidity caused by oxidation. (BHA is butylated hydroxyanisole. BHT is butylated hydroxytoluene. Both are synthetic chemicals — 100 percent man-made.)

Some nonfood materials are added to foods in order to make them more acceptable when they are dried. The most common is sulfur dioxide in dried fruits, especially those that are ordinarily light in color, such as apricots and peaches. Sulfur dioxide is GRAS, but may not be used in foods rich in vitamin B_1 since it destroys that vitamin. However, when used on fruits and vegetables, it preserves color and flavor and actually helps to retain vitamin C and carotene, which the body uses as vitamin A. One brand of prunes contains sorbic acid, which is also GRAS. Sorbic acid retards mold growth in the presence of high moisture. This brand contains more water than others and needs the sorbic acid for that reason. Not a bad idea: the prunes are moister to eat, and they cost less per pound, at least to the producer.

Dehydrated potatoes require several additives to maintain their quality. Sodium sulfite or sodium bisulfite, like sulfur dioxide, preserves color and flavor; BHT and BHA act as antioxidants to prevent fat in the product from becoming rancid; and citric acid (the acid found in lemons and oranges) helps the antioxidants to do their job more efficiently.

One class of foods preserved by drying that contains no additives at all is the legumes, such as peas, beans, lentils. We don't think of these as dehydrated vegetables, but they are the oldest and most widely accepted foods preserved by drying.

■ HEATING

Heating is a preserving method most of us are familiar with: canned foods are all preserved by heating. Unlike drying, which makes conditions unfavorable for microorganisms to grow, heating actually kills most bacteria, yeasts, and molds. Heating can sterilize a food completely if the temperature is high enough and the time long enough. (Heat preserving in its simplest form is *pasteurizing,* described under dairy products in chapter 3.)

Most microbes live at temperatures between 50° and 140° F. Below 50° they reproduce slowly and above 140° they begin to die.

When a food is acid, it takes less heat treatment to preserve it: the bacteria, yeasts, and molds die more quickly. (It's worth mentioning that there is very little chance of harmful microbes contaminating acid foods. Yeasts and molds are likely to contaminate, and although they generally spoil the taste and appearance, they are unlikely to be a hazard to health.) If a food is alkaline, more heat is necessary. That's why, when we preserve by canning at home, we have to use a pressure cooker for canning corn or beans, while we don't for canning fruits or tomatoes. Without pressure, we can only raise the temperature to 212°, but with pressure, we can heat foods to 240° or higher at home.

Next to sterilizing the food completely, the most severe heat treatment consists of raising the temperature of the food to its boiling point and then transferring it to sterile containers and sealing the containers. Jams, which are quite acid and contain only 35 percent water, are preserved in this way. Jams sometimes get moldy after they are opened, particularly if moisture gets into the jar, even if only on the surface through condensation. Condensation will form if the jar is removed from the refrigerator on a humid day, left open, and then closed. It is the high sugar and acid content of jams that helps to preserve them. The addition of water greatly dilutes sugar and acid and thus permits mold and yeast growth. Imitation jam (containing more than 35 percent water) requires a chemical preservative.

■ CANNING

This very important method of preserving foods was invented in 1813—before we even knew about bacteria! It was believed that "air" caused spoilage, and Nicolas Appert, the French inventor, processed with heat to get rid of the "air"! Not until Louis Pasteur and the great German bacteriologist, Robert Koch, did their work did we discover what really happened when we canned. And it wasn't until 100 years later that we learned that canning helped to preserve vitamin content by removing the air in the food to prevent oxidation.

Canning allows a form of heating that raises the food to a higher temperature than can be attained with ordinary heating, working in the same way that a pressure cooker does. Therefore, bacteria that would take much longer to kill by heating the food outside of a can at ordinary pressure may be more completely destroyed by heating within the can in less time. This heating is done after the food is placed in the can, and after the can is completely sealed, by placing the can in a pressure cooker where the temperatures can be raised to as much as 100° over the boiling point of water.

Commercial canning is quite different from home canning. Time and temperature control is extremely precise. For many years now, scientists have studied the behavior and life cycles of the microorganisms in canned foods and their reaction to canning conditions. Safety has been the most important objective. In all the years of commercial canning, there have been almost no cases of food poisoning from commercially canned foods. Most food poisoning cases are caused by improper home canning of nonacid foods.

Scientific studies have also shown how food may be canned so as to destroy its properties least. Canned food is cooked, and cooked for a long time. Texture and flavor are drastically changed, to the point where there is almost no resemblance to the fresh product. Because of these changes, additives are often used to compensate in some way. For example, sodium pyrophosphate in canned tuna, sodium EDTA in canned chick peas, and calcium chloride in canned tomatoes are added to prevent "watering," retain color, and keep firmness, respectively.

Two new heat-treatment methods of preservation are coming into popularity for some foods.

■ RETORTABLE POUCH

Aluminum-foil-pouch-packaged foods require much less heat treatment to sterilize the food, resulting in higher quality of taste and texture and probably requiring less compensation for changes caused by processing. This method is gaining great acceptance for consumer products in Japan and, so far, for a number of foods sold to restaurants in the United States.

The reason for the improvement is simple: heat penetrates a thin aluminum-foil pouch quicker than it does a metal can or glass jar, so the entire process can be completed in much less time.

■ ASEPTIC CANNING

This is another form of canning, which depends on keeping the microorganisms out of the food in the first place by operating under extraordinarily sanitary conditions, including the removal of germs from the air in the workroom itself with special filters. With this method, food may be sterilized at 300° F in less than one minute and is then put into sterilized cans. With less heat treatment required for sterilization, fewer additives are required to compensate for loss in taste, texture, and color.

■ FREEZING

This method of preservation has come closest to retaining the fresh quality of food when:

1. It is done right, that is, fast;
2. The food is not thawed and refrozen many times on its way from freezer plant to you;
3. The package is sealed tightly and doesn't dehydrate in your home freezer.

But all three conditions are rarely maintained and certainly cannot be depended on. Therefore, food processors often add materials to the food to try to keep it attractive and safe even if it does freeze and thaw many times.

As mentioned before, microorganisms grow slowly, or not at all, at very low temperatures: 0 to 10° F. Chemical reactions proceed very slowly at those low temperatures, as do enzyme-promoted changes, which is why freezing preserves.

But other changes take place. When water is frozen, it forms ice crystals. If the crystals are tiny, they do not puncture the cell walls of the food, and the food *texture* is preserved. However, if the crystals become large enough, they do break the cell walls and the texture breaks down. Some foods are more sensitive than others. Certain vegetables containing lots of water depend totally on the cells staying whole to maintain their texture — tomatoes, celery, lettuce, cucumbers — and cannot be successfully frozen at all. Tomato sauce can be frozen, but that's because we no longer care about the texture of the tomato in the sauce.

Even frozen baked foods become crumbly when the ice crystals become too large. Unfortunately, this can happen in many ways. Even if a food is frozen rapidly and the ice crystals are very small, simply storing the product at temperatures over 10° permits the crystals to start growing.

If the product is thawed in a car, then placed in your home freezer, it will then freeze slowly, and the crystals will become large.

Thawing frozen food too slowly can also form large crystals and degrade food texture.

Other things happen during distribution, storage, and thawing of frozen foods. For example, the surface of the food can be dehydrated, producing ugly "freezer burn"; or fluid can run out of the food, carrying with it flavorful components of the food, as well as drying it out.

The lack of control over all of these things gained a very bad reputation for frozen foods in the 1940s, which has taken years to overcome. Now, controls are much better, and methods have improved. One of the most important advances was the proper use of blanching before freezing, to prevent enzyme action in the product. More importantly, materials have been discovered that, when added to frozen foods, make them less vulnerable to unpleasant changes. The additives most commonly used in frozen foods are the antioxidants, which prevent rancidity, color change, and freezer burn. They are declared on the label when present.

Most frozen products contain no additives, especially the plain vegetables. But the more "prepared" the product is, that is, the more convenient for use, the more likely it is that additives will be present to protect against the changes caused by mishandling. (See chapter 11 for more information on frozen foods.)

Simple practical advice: get frozen foods home and into your freezer as fast as possible; follow thawing instructions exactly; and if you use only part of a package of frozen food, rewrap and pack it carefully in polyethylene bags to keep out moisture. (These are the common transparent bags that are sold for home use.)

□ CHEMICAL METHODS OF PRESERVATION □

Now we come to the dirty words: chemical preservatives. As we learned earlier, *everything* is a chemical, and every preservative that consists of any *material* is a chemical preservative. Vinegar is a fine chemical preservative consisting mainly of acetic acid and water. Salt is another.

Don't let the names throw you. All natural foods contain — indeed are — chemicals. The law requires that the ingredients of processed foods be listed, but not the components of the foods produced in the fields or as animals. "But," you may say, "that isn't what we really mean. It's the synthetic chemicals that are bad. You know: the chemicals made in factories."

The chemicals don't know how they are made. Pure salt, sodium chloride, is always sodium chloride, whether synthetic, mined, or evaporated and purified from seawater or your perspiration. Sodium propionate is always the same, whether produced synthetically or by bacteria while they are forming a cheese. Pure ascorbic acid is vitamin C, whether extracted from an orange, a tomato, or produced in a factory. The composition of vitamin C is identical in all sources, all will cure scurvy, and there is no way for the body or any scientist to tell the difference between pure ascorbic acid made synthetically and that found in nature. (Note: this means that paying extra money for natural vitamins is a waste. This is absolutely true.)

Some chemical preservatives have been with us for thousands of years: acids (vinegar), salt, sugar, wood smoke. The main chemical preservatives used today are:

Benzoate of soda (sodium benzoate)
Parabens
Sorbic acid and potassium sorbate
Propionic acid, calcium propionate, sodium propionate
Sulfur dioxide, and sodium and potassium sulfites
Acetic acid and acetates
Nitrates and nitrites
Diethyl pyrocarbonate (DEPC)

Antibiotics
Ethylene oxide and propylene oxide.

■ BENZOATE OF SODA (Sodium Benzoate)

Sodium benzoate is GRAS class 1. It is used mainly in foods that are acid, such as carbonated beverages, fruit juices, cider (to lengthen the shelf life by preventing yeast and mold growth, thus avoiding heating the product to preserve it), pickles, and sauerkraut. Please note that some brands contain this preservative and others do not. Its presence is *always* stated on the label. It is sometimes used in jams, jellies, preserves, margarine, and pie fillings. The more acid a food is, the less benzoate is required to preserve it. Citric acid often is added to foods that also contain sodium benzoate to improve the preservative effect of the benzoate.

Sodium benzoate is found naturally in cranberries, some plum varieties, cloves, and cinnamon. This should not necessarily be regarded as proof of its safety, for as we have mentioned before, any substance may be toxic if consumed in sufficient quantity. Early in this century, tests were conducted that showed that up to one-half gram of sodium benzoate per day, a large dose, was found not to be harmful, although some physiological changes were noted.

Diet sodas often contain sodium benzoate — a little less than 0.05 percent. If you drank a quart of diet soda a day, you would swallow that one-half gram. Watch cider labels, too. If the product is one that you might consume in large quantities, find a brand without sodium benzoate. You usually can.

■ THE PARABENS

These are first cousins to sodium benzoate and do the same thing. The name comes from *para*hydroxy*benz*oic acid, a derivative of benzoic acid. They work over a wider acidity range than sodium benzoate. The parabens have been tested extensively and have been shown to be toxic to animals at [illegible] levels, although tests done on dogs at levels of one gram per day for over a year showed the dogs to remain entirely healthy.

I can't think of a food where they are essential.

■ SORBIC ACID

This substance also is useful in the same types of products as benzoate of soda and the parabens. It is used as sorbic acid and as potassium sorbate. Either form is more effective than sodium benzoate and the parabens in preventing molds in margarine, cheese, bread, and cake.

It takes twice as much sorbic acid to kill a rat as does sodium benzoate. Of course, the levels of these substances used in foods are infinitesimal compared to the amounts used in testing — less than 1/100 of the test dosage. Sorbic acid is believed to be metabolized in a similar way to normal components of fat.

■ SALT

One of the oldest and most effective preservatives for many foods is salt. It reduces the amount of water available to the microscopic organisms and in that way slows down or stops their growth. When salt is used in combination with an acid and refrigeration, its effectiveness as a preservative is improved.

However, the amount of salt required to be effective as a preservative is much higher than is normally used for seasoning. For example, baked products, in which salt simply enhances the flavor without providing a salty taste, are about one-half of one percent salt. Savory-tasting products, rather than sweet products, are about one and one-half percent salt. By contrast, for preserving, five to ten percent of the brine is salt.

■ NITRATES AND NITRITES

The main purpose of nitrates and nitrites is to prevent the growth of that most dangerous of food-poisoning bacteria: botulinum. In addition, they improve the color of the meat.

As we discussed in "Food Additives and Cancer" in chapter 1, these substances have been associated with certain types of cancer, and their permitted level has been reduced.

■ SULFUR DIOXIDE AND SULFITES

Sulfur dioxide, made by burning sulfur, has a long record of use in dried fruits, fruit juices, and in wine making. It tends to hold back yeasts and molds as well as bacteria by forming sulfurous (*not* sulfuric) acid with the water in the food or juice. Sulfurous acid can also be obtained by dissolving sulfites in water.

There is no medical evidence of harm from sulfur dioxide. It forms compounds that are a normal part of our bloodstream and tissue — sulfates.

■ ACETIC ACID AND ACETATES

Vinegar is a naturally produced solution of approximately 5 percent acetic acid. Acetic acid may also be obtained by using sodium diacetate. Acetic acid is particularly effective in preventing the growth of salmonella and staph — two real villains in food poisoning.

Vinegar is the preservative (as well as a major seasoning) in ketchup, pickles, mayonnaise, salad dressing, and pickled fish. Years ago, bakers used a bit of vinegar to control molding of bread, and today some bakers use sodium diacetate to produce the acetic acid that does the same thing.

Sodium diacetate is made by combining acetic acid, the active component of vinegar, with sodium, to make it possible to add an acetic acid source in dry form to a food.

■ PROPIONIC ACID AND PROPIONATES

These are very effective in delaying mold formation, especially in baked foods. Propionic acid is naturally formed in the making of many cheeses, especially Swiss cheese. It is very commonly found throughout nature and is GRAS.

Propionic acid is usually added to the food in the form of either calcium propionate (in bread) or sodium propionate (in cakes). The amount used in baked foods is less than three-tenths of one percent (0.3 percent).

The safety data for the propionates is particularly favorable. Studies show that propionates are used like other fat substances in the body.

■ SUGAR

Whether brown, white, or molasses, sugar is a very effective preservative in candy, jellies, jams. It takes about a 65-percent solution of sugar to be an effective preservative against the growth of yeasts and molds and most bacteria (especially if the food is also acid, like all fruit products).

Sugar acts as a preservative by preventing the microbes from using the water available, in the same way that salt does, except it takes ten times as much sugar to do so.

■ WOOD SMOKE

This very "old-fashioned," "natural" preservative works only because when the wood is burned, chemicals are produced that are very effective. Those chemicals are *formaldehyde, carbolic acid,* and a host of others. Because of environmental restrictions, the direct smoking of meat is declining. Instead, the wood smoke is produced in a factory, under sealed conditions, condensed into a liquid, and sprayed or painted onto the meat. This eliminates the preservative effect of smoking, except on the surface, and most of the smoky flavor is lost.

In the old-fashioned smoking method, not only were the chemicals impregnated into the meat by the smoke, but the surface of the meat was also dried out some, which also helped to preserve; and, of course, the meat was heated, which killed many of the microorganisms present.

Many carcinogens have been identified in wood smoke (as well as in cigarette smoke), and hence moderation is advised in eating smoked meat.

■ HYDROGEN PEROXIDE

This chemical is used to treat milk that is used to make Swiss cheese. It kills certain harmful bacteria while leaving intact the bacteria needed to produce the cheese. Hydrogen peroxide will not be found on the cheese label since its use in the process is permitted under the cheese standards without being mentioned on the label.

While there are a few other preservatives, they are rarely used. As we run into them on the labels we look at, they will be explained.

Chemical preservatives are not so frightening after all; some of them are quite natural, although that doesn't necessarily mean a thing as far as safety is concerned. *None* of the preservatives that are permitted in foods are known to be *significantly* harmful. The nitrates and nitrites undoubtedly will be replaced in time.

If there were no other dietary restrictions, I would feel free to eat foods containing sorbic acid, acetic acid, proprionic acid, sugar, sulfur dioxide, and salt, and I would be moderate in my use of foods containing sodium benzoate, parabens, nitrites, and those foods that have been smoked. Those people with high blood pressure should keep their salt consumption down, and diabetics must restrict their sugar intake.

There seems to be no justifiable reason for fear of "chemical preservatives" in general.

□ PRESERVATION WITH MICROORGANISMS □

As mentioned earlier, all the microorganisms do is manufacture chemical preservatives right in the food itself. Sour pickles are preserved with salt and lactic acid; the lactic acid is produced by bacteria. This is also true of yogurt and many of the cheeses. The two most common types of bacterial preservation are (1) the lactic-acid-producing bacteria and (2) the acetic-acid-producing bacteria.

In the most common type of preservation using yeast (also a microorganism), the yeast promotes the production of alcohol, which is a very effective chemical preservative.

The principal foods preserved through bacterial action are sour pickles, wine, sauerkraut, and cheeses.

The choice of a physical method or chemical method for preservation is made by the food processor to produce a product acceptable to consumers. As you will see, some processors use preservatives in certain products that other producers do not. This can be a result of better sanitary practices by those who do not, or differences in opinion about the need for the preservative. Remember also that some products have preservatives added to protect them *after* you have opened the package and have no purpose before that. They are there to protect the food against contamination by you, in your home.

Chapter 3

NUTRITIONAL CHEMICAL ADDITIVES

The Vital 21

In the nutritional scheme of things, we need carbohydrates and fats to provide calories (the fuel needed by our bodies), plus twenty-one other substances for healthy functioning; indeed, for the maintenance of life itself.

It is not possible for the human body to produce new cells and muscle tissue and to sustain life without protein, one of the twenty-one. We require from 1½ to 2 ounces (45 to 65 grams) of protein per day. The other twenty materials — the vitamins and minerals — are required in very small amounts, but exert a profound influence on health. To put it more forcefully, the absence of any one of them causes severe illness in the form of deficiency disease.

Since there are many good books about vitamins and minerals, if you are interested in learning about them in depth, I will not go into too much detail here, but I will present very briefly the main sources of each as well as some key things to remember when selecting foods.

In 1980, the U.S. Food and Drug Administration took a major step to assure us of superior nutritional food quality. After many years of study, consumer surveys, discussion with interested groups of scientists, industry representatives, and consumers, a policy for nutritional fortification of foods was issued. (See appendix 3.) The highlights of this policy are:

1. Nutrient addition can be an effective way to improve the overall food supply.
2. Snack foods, such as candy and carbonated beverages, should not be fortified: it encourages people to believe that such foods are nutritionally adequate.
3. Fresh foods, such as meat, poultry, fish, and fruits and vegetables, should not be fortified.
4. If foods are fortified to restore nutrients lost in storage or processing, they should be fortified with all the nutrients lost in processing, not just some.
5. Until recently, fortification levels in foods were based on a calorie intake of 2800 per day in the average diet. New studies show that our average intake is now 2000 calories. Therefore, producers of fortified foods are being encouraged to fortify so that there is a greater balance among daily nutrients required at this 2000-calorie level. This means that the concentration of vitamins, minerals, and protein will be higher in foods such as fortified ready-to-eat breakfast cereals. In simple terms, this means more vitamins and minerals in less food.
6. If a food is fortified to provide "balanced nutrients," it should provide all of the twenty-one in balance as shown in the chart on page 52.
7. So that consumers will not be misled, any food that claims to contain balanced nutrients should provide at least 40 calories in a single normal serving, that is, not less than 2 percent of the day's calories. That serving should also contain not less than 2 percent of the day's nutrients, balanced. This will discourage the marketing of so-called balanced-nutrients foods that are really inconsequential nutritionally. For example, a small piece of hard candy, fortified with nutrients, contains only 12 calories, thus supplying less than 1 percent of the day's calories, and less than 1 percent of the nutrients. The FDA doesn't want consumers misled by claims of "balanced nutrients," when actually the amount of food is grossly insignificant compared to the body's total daily needs.
8. Label claims that can mislead consumers are more clearly prohibited. For example, an imitation food cannot be

□ NUTRIENT REQUIREMENTS □

NUTRIENT	UNIT OF MEASUREMENT	U.S. RDA[1]	AMOUNT/100 CALORIES
Protein (optional)[2]	gram (g)	65 45	3.25 2.25
Vitamin A	international unit (IU)	5000	250
Vitamin C	milligram (mg)	60	3
Thiamine (B_1)	milligram (mg)	1.5	0.075
Riboflavin (B_2)	milligram (mg)	1.7	0.085
Niacin	milligram (mg)	20	1.0
Calcium	gram (g)	1	0.05
Iron	milligram (mg)	18	0.9
Vitamin D (optional)[2]	international unit (IU)	400	20
Vitamin E	international unit (IU)	30	1.5
Vitamin B_6 (Pyridoxine)	milligram (mg)	2	0.1
Folic Acid	milligram (mg)	0.4	0.02
Vitamin B_{12} (Cobalamin)	microgram (mcg)	6	0.3
Phosphorus	gram (g)	1	0.05
Iodine (optional)[2]	microgram (mcg)	150	7.5
Magnesium	milligram (mg)	400	20
Zinc	milligram (mg)	15	0.75
Copper	milligram (mg)	2	0.1
Biotin	milligram (mg)	0.3	0.015
Pantothenic Acid	milligram (mg)	10	0.5
Potassium	gram (g)	2.5[3]	0.125
Manganese	milligram (mg)	4.0[3]	0.2

[1] Recommended Daily Allowance

[2] Fortification with these nutrients is at the option of the food manufacturer in balanced-nutrient foods.

[3] Not a U.S. RDA, but recommended by the Food and Nutrition Board, National Academy of Sciences–National Research Council.

These RDAs are FDA-established. They are subject to change with new knowledge. (Adapted from *Federal Register 45*, no. 18, p. 6323) claimed to be superior to a natural food simply because it has vitamins added to it.

9. There must be reasonable assurance that excessive nutrient consumption will be avoided, taking into consideration the amounts of the nutrient likely to come from other dietary sources as well.
10. Steps to protect against overdoses of nutrients will be provided.

These policies will have an influence on labeling and our food supply for years to come.

□ THE VITAMINS □

A simple way to think about vitamins is to remember three things about them:

1. Most cannot be produced by the body, which means they must be supplied by foods.
2. Their absence produces disease.
3. The amounts required are extremely small, less than one-half gram per day for *all* our vitamin requirements combined. A few thousandths of a gram of each vitamin is all we require daily.

The vitamins and minerals are required to assure proper metabolism in the cells. Some act like hormones, some as catalysts (which promote changes), some to make enzymes, and in the case of minerals, to form body substances that are required for life — for example, iron is needed to make hemoglobin, the red-blood-cell material.

The RDAs (Recommended Daily Allowances) listed on the chart are greater than the amounts needed merely to prevent diseases resulting from deficiency; rather, they represent amounts judged to be required for good health.

Since there is a great deal of debate about megavitamin doses — amounts many times in excess of the RDA — it helps to know a bit more about the nature of each of the vitamins. There are basically two types of vitamins: those that are *water soluble* and those that are *fat soluble.* The water-soluble vitamins cannot be stored in the body to any degree since they move with water through and out of the body if not needed. The fat-soluble vitamins, however, can be stored, since our body stores fat; for this reason, there can be such a thing as too much of these vitamins.

THE FAT-SOLUBLE VITAMINS (Vitamins A, D, E, and K)

FAT-SOLUBLE VITAMINS

	Function	Deficiency Results
Vitamin A	maintains function of epithelial cells, mucous membranes, skin, bone; constituent of visual pigments	night blindness, glare blindness, rough, dry skin, dry mucous membranes, xerophthalmia
Vitamin D	calcium and phosphorus absorption and use in bone growth	rickets, soft bones, bowed legs, poor teeth, skeletal deformities
Vitamin E	protects cell structures	red cells hemolyze more
Vitamin K	for blood clotting	slow clotting time

Vitamin A

Main sources: Liver, eggs, yellow and green vegetables. A small serving of carrots or spinach or sweet potato supplies more than a day's requirement in a form called carotene, which is converted to vitamin A by the body. Other foods that supply vitamin A are milk and butter.

Vitamin A is added to many foods in one of three forms: retinol, retinol acetate, and retinol palmitate. All forms occur in nature and are also made synthetically. All are GRAS. At one time, vitamin A was extracted from fish-liver oils and was used in the early days of food fortification, but there was a slightly unpleasant fishy taste. Refinements in extraction methods and later synthetic production of vitamin A greatly reduced this problem.

All three forms are equally valuable nutritionally. They vary in stability and solubility, and the form is chosen on the basis of which is best for the food being fortified, to make sure it keeps its potency.

Breakfast foods and low- and nonfat milk are fortified with vitamin A, and the labels clearly state the source and percentage of RDA supplied.

In the United States, there is very little likelihood of vitamin-A deficiency in a diet that includes liver or fruits and vegetables, and a reasonable amount of milk, either whole or low-fat, since the low-fat milk is fortified up to the level of the full-fat milk.

The scientific literature warns that excessive vitamin A is undesirable and harmful, with symptoms such as loss of appetite, bone pain (especially in the long bones), fragile bones, and loss of hair.

The addition of vitamin A as a supplement to a diet supplying the RDA of 5000 IU is not recommended. In older people especially, a dietary supplement may be necessary, but this should be held below the level of 10,000 IU per day.

Beware of megadose vitamins!

Vitamin D

Main sources: Sunshine on the skin. There is no significant normal food source, although fish-liver oils contain large amounts.

We are dependent on fortified foods for sufficient vitamin D, especially if we live in a less sunny northern climate, in cities, or indoors.

The pigment of dark-skinned people protects them from

getting too much vitamin D when in the sunnier tropical areas, but this tropical protection becomes a problem in the temperate zones, where the very same pigmentation may prevent sufficient vitamin-D formation.

The principal food fortified with vitamin D is milk, which is fortified so that one quart provides 400 IU, the RDA. A number of breakfast cereals are also fortified so that one serving provides 10 percent of the RDA.

Too much vitamin D produces high blood sugar, nausea, and kidney stones, and may even be fatal, but there is no way to obtain excessive vitamin D from natural sources or even fortified foods; excessive amounts can only be obtained from improper use of pharmaceutical preparations.

Labeling: There are two forms of the vitamin used for fortification: vitamin D_2 and vitamin D_3. They are so declared when added. The vitamin D_2 is made by irradiating a substance called ergosterol with ultraviolet light. The D_3 is separated from fish-liver oils. There is no reason for consumers to discriminate between the two. The choice is made by the manufacturer and is based on physical properties and taste of the food product. Both are effective in preventing deficiency diseases and promoting good health.

Vitamin E

Main sources: Vegetable fats—corn, cotton, soybean oils. Vitamin E is easily obtained, since it is present in most oils and cereal grains.

The RDA is 30 IU—and no deficiency has ever been reported! It is only recently that the need for vitamin E in human nutrition was recognized and an RDA established. While there is some popular literature pushing vitamin E for greater endurance and prevention of heart disease, there is no scientific evidence (that is, no evidence published in scientific journals, and repeated by other scientists, and confirmed) nor agreement by many scientists that this is the case. On the other hand, there is no evidence of harmfulness through high intake either.

It is known that vitamin E is important in the prevention of spontaneous abortions in pregnant rats, and a deficiency causes the loss of reproductive ability in male rats, but no such connections have been found for humans.

In humans, laboratory tests show that the result of a deficiency is a tendency of certain blood cells to disintegrate. Also, the amount of vitamin E needed increases with higher polyunsaturated-fat consumption.

Labeling: Vitamin E is added to some foods as a tocopherol mixture, or tocopherol acetate, or alpha tocopherol — three active forms of vitamin E.

The tocopherols are extracted from vegetable oils, such as wheat-germ oil. The acetate is made using the extracted tocopherols, and is much more stable. As used in foods, there is no difference in vitamin value.

Summary of vitamin E: just forget about it. You get enough in any normal diet.

Vitamin K

This is one vitamin you don't hear much about and will not find used as an additive. It affects the clotting time of blood and prevents hemorrhages. The average diet supplies four to eight times the estimated requirement.

A fact for your interest: about half of the body's requirement for this vitamin is produced by bacteria in your intestinal system. If for some reason there should be a lower than usual vitamin-K level in the diet (while you are taking antibiotics over an extended period, for instance), it would be possible to produce a deficiency, which would increase blood-clotting time.

Vitamin K is sometimes deliberately prescribed by physicians to reduce blood-clotting time.

■ THE WATER-SOLUBLE VITAMINS

The supply of this group of vitamins needs to be replenished about every day, since any surplus consumed is simply excreted. Actually, there is a slight storage capacity for these vitamins, but the only practical way to ensure adequate consumption of them is to include them daily in your diet.

□ THE WATER-SOLUBLE VITAMINS □

	Function	Deficiency Results
Vitamin C (ascorbic acid)	cell oxidation-reduction balance	scurvy, sore mouth, sore and bleeding gums, weak-walled capillaries
Thiamine	coenzyme; helps carbohydrate metabolism	beri-beri, poor appetite, fatigue, constipation
Riboflavin	coenzyme; carbohydrate, fat, and protein metabolism	eye sensitivity, cheilosis
Niacin	coenzyme; carbohydrate, fat, and protein metabolism	pellagra, dermatitis, nervous depression, diarrhea
Vitamin B_6	coenzyme; amino-acid metabolism	convulsions, anemia, kidney stones
Folic Acid	blood formation, synthesis of DNA, RNA, and choline by the body, amino-acid metabolism	megaloblastic anemia, glossitis, diarrhea
Vitamin B_{12}	growth, blood formation, choline synthesis, amino-acid metabolism	macrocytic anemia, sprue, pernicious anemia
Pantothenic Acid	coenzyme; carbohydrate, fat, and protein metabolism	unknown
Biotin	fatty-acid synthesis, other body chemistry reactions	lassitude, anorexia, depression, anemia

Vitamin C

Main sources: Citrus fruits, raw leafy vegetables (lettuce, cabbage), berries, tomatoes. Vitamin C is added to many processed drinks and imitation juices to bring the level close or even equal to that of orange juice.

Vitamin C is one of the most controversial vitamins at present. One theory is that massive doses — twenty to fifty times the RDA — will prevent or stop colds. There is no real proof that it is effective in such doses, and there is some concern that such great doses may produce undesirable side effects, though this has not been proved either. Extensive research is being done to learn more about massive daily intake of vitamin C.

It is quite easy not to get enough vitamin C in your diet unless you deliberately eat citrus fruits, tomato juice, or one of the fortified products. It doesn't take a great volume to provide 100 percent of the RDA — 4 to 5 ounces per day of any form of orange juice will do it, even if you have no other source in the diet. It takes four times that amount of tomato juice (16 ounces) to supply the RDA.

Labeling: Ascorbic acid is shown in the ingredient listing. Don't confuse this with *sorbic* acid, which is unrelated and is used to stop mold growth. Ascorbic acid (or sodium ascorbate) is vitamin C, and is produced synthetically using a fermentation process.

You may see ascorbic acid listed among the ingredients of some cured meats. The purpose there is not for vitamin fortification; ascorbic acid is believed to reduce the formation of harmful materials (nitrosamines) when treating meat with sodium nitrite. It also helps produce a more attractive color in the cured meat.

The widespread use of vitamin C as a fortifying nutritional additive is possible because it is produced synthetically at a very low cost. The cost of fortification is negligible.

The B Vitamins

○ Thiamine (Vitamin B_1) ○

Main sources: Unlike some other vitamins, there is no one single food that provides 100 percent of the RDA in a single serving. The daily amount needed must be obtained from a variety of foods: pork, lean meat, grains, green vegetables. Fruits and milk are not sources of thiamine.

As shown in the chapter on baked foods, enriched bread is as high in thiamine as whole-wheat bread. There are many other processed foods available that are fortified with this vitamin. This is particularly true of breakfast cereals. The nonsense about food processors robbing grain of its vitamins and barely putting some back is just that — pure nonsense. The natural cereals simply don't provide more than 10 percent of the RDA in a serving, while the same size serving of many of the processed cereals provides 100 percent!

Mind you, I'm not saying, "Don't eat natural foods," only, "Don't be misled into thinking that natural grain products provide more vitamin B_1 (thiamine) than the fortified cereals."

Labeling: There are two forms of synthetic thiamine used as supplements: thiamine hydrochloride and thiamine mononitrate. The mononitrate is more stable in foods than the hydrochloride, was developed later, and is more likely to be used. There is no difference between them of any consequence to the consumer, thus no reason for choosing one over the other. As far as consumers are concerned, the vitamin potency must be protected by the manufacturer, and a surplus is added so that the user is assured of receiving the amount claimed.

This vitamin is listed as thiamine in the RDA table.

○ Riboflavin (Vitamin B_2) ○

Main source: Milk, whole or skim; liver; kidneys. One glass of milk provides 25 percent of the RDA, but like thiamine, there is no one food that supplies 100 percent of the RDA in a normal portion. It is also found in grains, but in low amounts, except for wheat germ. However, it would take about a quarter pound of wheat germ to supply 100 percent of the RDA.

The foods that are fortified with thiamine are usually fortified with riboflavin as well, with the breakfast cereals being the most prominent group. It is probably a good idea to include such fortified foods in your diet or use a vitamin capsule.

Lack of riboflavin causes eyestrain and eye itching and burning. (I've been stopping sties in my kids' eyes with riboflavin for years: at the first sign of a sty, we take a vitamin pill containing riboflavin for a few days and they rapidly disappear. This anecdote may support the view that additional vitamins are helpful under stress conditions.)

Labeling: Riboflavin is used in two forms: riboflavin and riboflavin phosphate. (The phosphate is more soluble and convenient for food fortification.) They may be considered as identical in value. This vitamin is listed as riboflavin on the RDA statement.

○ Niacin ○

Main sources: Liver, fish, chicken. One portion provides one-third to one-half of the RDA. Other sources are the enriched or fortified foods. Since we began enriching bread with niacin, the deficiency disease pellagra has been all but wiped out in the United States.

The accurate chemical name for niacin is nicotinic acid, but in order to avoid confusion with nicotine, to which niacin has no resemblance or connection, its name was changed to niacin.

Labeling: This vitamin is made synthetically and is used to fortify foods in one of two forms: niacin and niacinamide. While therapeutic doses of niacin may cause hot flashes, this is not true of niacinamide. Needless to say, there are no therapeutic doses in foods, whether fortified or not.

○ Vitamin B_6 (Pyridoxine) ○

Main sources: Pork, liver, lamb, veal, potatoes, legumes, wheat germ, oatmeal. The amount we require depends on the amount of protein we eat. The 2 milligrams RDA will take care of 100 grams of protein per day, which is a great deal.

Pyridoxine is destroyed by heat. The importance of vitamin B_6 in human nutrition was discovered when the vitamin was accidentally destroyed in a canned-milk formula for infants, resulting in nervous irritability and convulsions, which were immediately stopped upon injection of vitamin B_6, proving that the symptoms were the result of a deficiency.

As with so many other B vitamins, no one food will supply the RDA in a single portion, and fortification of foods with this vitamin is common.

Labeling: Pyridoxine is produced synthetically and is used in only one form, labeled as pyridoxine hydrochloride in the ingredients statement, and as vitamin B_6 in the RDA table.

○ Folic Acid ○

Main sources: Folic acid is not found in foods as such, but folates are found that are converted to folic acid by the body. Green leaves (foliage), liver, meat, and fish are sources of the raw material needed.

The RDA for this vitamin is 400 micrograms (*not* milligrams) per day. That's four-tenths of one milligram.

While some foods are fortified with folic acid, it is not GRAS and its use is limited.

Folic acid plays a crucial role in preventing anemia through its role in blood formation, and is involved in the treatment of pernicious anemia. There is a danger that high intake of folic acid may cure the anemia, so that the blood examination looks normal, but folic acid alone is not enough to stop the neurological symptoms of pernicious anemia. Therefore its use might prevent the proper recognition and treatment of the complete disease!

Labeling: When used in fortification, it is labeled as folic acid or folacin — two names for the same substance.

○ Vitamin B_{12} (Cobalamin) ○

Main sources: Seafood, meat, eggs, dairy products. Not available from vegetables. Some breakfast cereals are fortified with vitamin B_{12}. The RDA is 6 micrograms. Pure vegetarian diets can result in vitamin B_{12} deficiency and the pernicious anemia caused by this deficiency.

This vitamin is produced commercially using microorganisms: we literally farm bacteria specially selected and grown to produce vitamin B_{12}, which we then isolate and use. It is not made synthetically.

Vitamin B_{12} cannot work without another substance, called the "intrinsic factor," produced in the lining of the stomach. Administering vitamin B_{12} by injection will cure the neurological symptoms of pernicious anemia, but the intrinsic factor is needed for the vitamin to be absorbed from the food supply.

It took twenty years after the discovery that liver extract could treat pernicious anemia to identify the B_{12} factor in liver that did the job, and then another twenty to learn its chemical structure.

Another curious fact about vitamin B_{12} is its enormous potency—it is one of the most potent substances known. As little as 6 micrograms, an amount that is *one-millionth* of a pound, an amount you can barely see, a crystal or two, is enough to produce a healing effect in a person suffering from a deficiency.

Labeling: It is listed as cobalamin or cyanocobalamin in the ingredient legend and as vitamin B_{12} in the RDA statement.

○ Pantothenic Acid ○

There is no RDA for this vitamin, but the amount considered safe and sufficient is 4 to 7 milligrams per day.

Main sources: Liver, kidneys, heart, eggs, and, to a lesser degree, broccoli, mushrooms, and cereals. The average diet provides 5 to 10 milligrams per day—well beyond the estimated need.

Labeling: When used to fortify foods or as a vitamin supplement, calcium pantothenate or d-pantothenyl alcohol is used. Both are GRAS and are made synthetically.

○ Biotin ○

There is no RDA for this vitamin, but rather an estimated daily adequate amount of 0.1 to 0.2 milligrams. The average diet provides more than this amount. Liver, mushrooms, peanuts are particularly high in biotin. It is also produced by bacteria in the digestive system.

Raw egg white destroys this vitamin, and a diet high in raw egg white can produce a deficiency.

Biotin is produced synthetically for use in foods or vitamin supplements. It is GRAS.

○ Other Vitamin-B-Complex Factors ○

Para-aminobenzoic acid (PABA): This is no longer considered a vitamin.

Choline: This is not yet confirmed as a vitamin but is produced and sometimes used to fortify foods, although it is made by the body itself and is widespread in foods. There is no evidence of problems caused by deficiency. Two forms are used: choline bitartrate and choline chloride.

Inositol: This was thought to be a vitamin back in 1940, but is no longer considered as such. Apparently we can produce all we need in the body.

□ MINERALS □

Like vitamins, small amounts of minerals are required to maintain health and to avoid disease. Fortunately, our diets usually contain sufficient minerals to meet our requirements, with a few notable exceptions. The exceptions are worth learning about.

■ CALCIUM AND PHOSPHORUS

These two minerals are required in fairly large amounts daily: nearly 1 gram of each (800 mg). They are extremely important, especially in young and old people, for bone development and maintenance of bone strength.

Main sources: If the diet normally contains dairy products, mainly milk in any form, and at least two 8-ounce glasses of milk per day are consumed, there will be enough calcium and phosphorus in the diet.

But many folks do not drink milk, and those people must

find some way of getting more calcium and phosphorus into their diets. Cheddar cheese—*but not cottage cheese*—is a good source of calcium, as is enriched bread. Whole-grain cereals provide very little calcium.

Simple rule of thumb: one cup of milk will supply one third of the daily requirements, four slices of enriched bread, another 10 percent. One slice of American cheese equals one glass of milk. Sardines and salmon are also high in calcium and phosphorus.

However, it is easy enough to take a calcium and phosphorus supplement, with no risk of an excess, if you're worried that your diet does not provide enough of these minerals.

■ IRON

There is a more widespread deficiency of iron than of any other mineral in the American diet. Anemia caused by lack of iron is common, especially among women under 50 years old.

The RDA is 10 milligrams for adult males, 18 milligrams for females under the age of fifty.

Main sources: Calf or lamb liver, meat, fish, poultry, eggs. Liver supplies as much as 14 milligrams in a 4-ounce serving, while these other foods provide 2 to 4 milligrams in a 4-ounce serving.

Whole-wheat bread and enriched white bread each contain the same amount of iron, with a good possibility that the form of iron in the white bread is more readily absorbed than that in the whole wheat.

Iron enrichment of food is necessary and common. The forms of iron used are many: reduced iron (actually pure iron), ferric phosphate, ferric pyrophosphate, ferric sodium pyrophosphate, ferrous gluconate, ferrous lactate, ferrous sulfate. All of these forms of iron are GRAS.

The food processor will use the form that he thinks is the best combination of cost, flavor effect, stability, and effect on stability of the food, effect on color, texture, and so forth. As consumers, we should pay no attention whatsoever to the iron source (with the exception of ferric oxide, a class 5 substance), but only to the amount and percentage of the RDA on the label.

There are many iron supplements available for those whose diets are deficient in iron. Excessive iron has no harmful effect, but stools may turn black with some forms.

■ IODINE

The RDA for iodine is minute: 0.15 milligram per day — otherwise called 150 micrograms.

Iodine deficiency causes goiter (impaired thyroid-gland function, eyes bulging), which is immediately cured by adding iodine to the diet. Goiter is often linked to geography — that is, it appears where the soil is deficient in iodine or where no seafood is available (seawater contains iodine). Goiter as a result of dietary iodine deficiency has been reduced substantially through adding iodine to salt (iodized salt). If you consume seafood or iodized salt, you need not be concerned about iodine deficiency in your diet.

Main sources: Fish, iodized salt, vegetables grown in high-iodine soil near seacoasts.

When iodine is added to salt it is added either as cuprous iodide, which is GRAS but only permitted in salt up to .01 percent; or as potassium iodide, which is also GRAS. When potassium iodide is used, there is a strict limitation on the amount that may be added to a food to supply iodine, so that no more than a total of 0.225 milligrams of iodine will be consumed in a day by adults, with less for children.

■ ZINC

In recent years, the importance of zinc in the diet has been learned. The RDA is 15 milligrams. Zinc deficiency slows down wound healing, produces anemia, and slows growth in children. Alcoholism results in zinc deficiency.

One study of 150 children from middle-income families

showed 8 percent had a zinc deficiency. We need to be aware of our zinc requirements.

Main sources: Beef, liver, and oysters (which contain about 25 milligrams per oyster!). Wheat germ contains 5 milligrams per ounce. No one food, other than oysters (Atlantic only), can supply the day's requirement in one portion, but zinc does seem to be widespread enough so that a balanced diet will supply the day's requirements.

A number of foods, especially breakfast cereals, are fortified with zinc.

Labeling: The forms in which zinc are added to foods are zinc chloride, zinc gluconate, zinc oxide, zinc stearate, and zinc sulfate.

Other minerals that are required are magnesium, sodium, sulfur, potassium, chloride, copper, manganese, fluoride, selenium, cobalt, and chromium.

Manganese is sometimes added, as are magnesium and potassium. Sodium and chloride are amply supplied; in many cases, too amply supplied in the form of table salt. The other minerals are micronutrients, trace elements found in the normal diet.

□ LABELING SUMMARY □

Many of the frightening chemical names on food labels are the names of the food supplements added to improve the nutrition of the food products. Control over the amounts that may be added is very strict. The nutritional value of the added nutrients is identical to that of the same nutrients found naturally. You can use the nutritional label information on processed foods to assure yourself of getting foods with sufficient vitamins and minerals.

Chapter 4

COLORS AND FLAVORS

□ COLORS □

Colors are added to foods to enhance or change their appearance to increase their consumer appeal. In some cases, there would probably be no market for the product without added color. Maraschino cherries would be a pallid yellow; cherry soda would look like water. In other cases, color is added to make the food look more expensive than it really is, such as the use of yellow color in bakery foods to suggest a higher content of egg yolks.

Sometimes color is added because our basic conception of the food is based on a color, such as the yellow in butter. White butter might be unappealing, so most butter is colored yellow. Margarine was a nothing product until it was permitted to be made the color of butter. All margarine is now colored.

Many formulated and processed foods are colored to provide an attractive appearance—such as gelatin desserts. They would all be an ugly, unappetizing color without color additives. Color is sometimes added because the manufacturer wants the product to have more eye appeal than the competition's product.

Another reason to add color is to make processed foods *uniform* in color. Nature does not produce uniformly colored products, but consumers expect the color to be uniform when they buy a processed food. Otherwise, they might suspect that the product has been changed.

Two kinds of coloring materials are used: dyes and natural colors. The dyes are synthetically produced and go through a rigorous testing and inspection procedure by the FDA to obtain certification. Every *batch* of dye used in foods must be certified by the FDA. After the first Food and Drug Act was passed in 1906, only seven dyes were permitted, reduced from the more than seventy that were in actual use at the time. Over a period of time, some colors were added. As safety-testing methods became more sophisticated and were applied to these colors, some were no longer permitted.

Many that survived early safety-testing procedures were outlawed later as a result of the passage of the Color Additives Amendment of 1960, which also includes the Delaney Clause.

■ TESTING

At present, there are eleven dyes that are permitted in foods, all of which must be certified. Of these, five are provisionally permitted. These are the "F, D, and C" colors (from the *F*ood, *D*rug and *C*osmetic Act of 1938). The certification procedure is designed to assure purity of composition rather than to determine safety. It is meant to ensure that the dyes that have already been tested for safety are produced to meet standards of purity for food use.

In addition to the dyes, there are many other color additives that are "natural" colors and do not require certification: turmeric, oleoresin paprika, paprika, carotene, caramel, and beet-powder extract.

There are also some very "unnatural" colors that do not require certification: titanium dioxide and ultramarine blue are two. Most white products that contain artificial color use titanium dioxide as the coloring material.

Whether foods are colored with certified color or natural color, they still must be labeled as "artificially colored." However, in some foods where oleoresin paprika, paprika, or turmeric is used, they are simply declared as ingredients without indicating that the purpose is for color. They are present as spices, and these spices also color.

Some standardized foods permit the use of color without any label declaration. This is true of the standardized cheeses, butter and — most important — artificial color in ice cream, although declaration is required in other frozen desserts such as sherbet, ice milk, and water ice. The FDA is seeking to change this regulation.

□ FLAVORS □

"Natural flavors," "natural flavoring," "artificially flavored," "artificial flavor," "natural and artificial flavor" . . . Those are the words we will find on many food labels. What do they mean? What *is* a flavor, as the word is used to state an ingredient of a food product?

Flavors are the substances that impart taste and aroma to a food. The sense of taste itself detects only sweet, sour, salty, and bitter, or a blend of those. Only in its aroma, detected by the sense of smell, does a food become truly unique in its flavor.

To see clearly and dramatically the difference between taste and aroma, hold your nose and chew a piece of apple. Then repeat the same process, this time with a piece of onion. Be sure to close your nostrils tightly while you do this. All you will taste is sweet — in both cases. Only the addition of the sense of smell enables you to distinguish the flavor difference between the onion and the apple.

The flavors present in foods such as fruits and spices can be separated from the food and used to impart the flavor to something else. The flavor is carried in what is called the essential oil, so named because it contains the flavor essence characteristic of the food. For example, the flavor of nutmeg, in the form of oil of nutmeg, can be separated from the nutmeg and added to coffee-cake dough to delicately flavor it. Similarly, essential oils of orange, lemon, grapefruit, and many other fruits can be separated and used for flavoring purposes.

Why do processors bother to do such a thing? Some reasons

are: as a means of having the flavor available long after the source of the oil has spoiled; and to provide a source of uniform flavor, much more so than the original spice or fruit, and also in a very much more stable form.

These flavors are not pure chemical substances, but are a blend of many substances. We had to learn what the precise composition was that produced such strange and wonderful smells and tastes. During the nineteenth century, chemists were extraordinarily successful in analyzing the components of many natural flavors and, even more significant, learned to synthesize many of them. For example, extract of vanilla contains vanillin, the principal component of vanilla flavor; chemists soon were synthesizing and using vanillin as a flavor instead of the vanilla extract. Unfortunately, vanillin is not the only substance in vanilla flavor, and we still don't have a synthetic duplication that tastes like the real thing. But synthetic vanillin is a precise duplicate of the vanillin found in vanilla extract from the vanilla bean.

In some cases, the result has been ridiculous. Methyl anthranilate is the principal material responsible for typical Concord grape flavor. It is widely used as an artificial flavor in grape drinks, sodas, and ice confections. Methyl anthranilate is not the only component of grape flavor, and it has a very strong, coarse grape character. Children actually prefer grape drinks flavored with the synthetic anthranilate to the real thing, it is so much less subtle.

The labeling of flavors is not helpful enough to the consumer. The mere word "artificial" tells us nothing since the substance may be a synthetic duplicate of a component of natural flavor. Chances are when the label says "artificial," the overall taste sensation will not be as smooth as a naturally flavored product, but this is not inevitably so at all.

There are over *seven hundred* synthetic flavors permitted as food additives that are not GRAS. There are twenty-four synthetic flavors that are GRAS. The consumer has no way of telling which is in the food: the GRAS product or the non–GRAS additive.

Do not assume that because an extract is natural that it is safe. One of the most dangerous materials is good old-fashioned oil of sassafras, out of which root beer was once made. It turned out that a natural substance in oil of sassafras, safrole, is a rather nasty cancer producer. This material is no longer permitted in foods; only safrole-free sassafras extract may now be used.

There is an enormous amount of ignorance on the subject of flavors. One theory presented is that *all* additives and especially artificial flavors cause hyperactivity in children, but nobody has been able to duplicate the work reported. On the other hand, nobody has ever claimed that highly flavored, artificially flavored foods are good for you, and on the general principle of moderation, especially in the presence of ignorance, it is probably a good idea to avoid eating and feeding your children a great deal of artificially flavored, or even strong naturally flavored, foods or drinks.

■ FLAVOR ENHANCERS

Some materials are used to enhance or intensify the flavor of foods. MSG (monosodium glutamate), disodium guanylate, and disodium inosinate are such substances. MSG is GRAS, and the others are additives.

MSG is considered to be responsible for the "Chinese restaurant syndrome" (a headache after the meal), so called because many Chinese restaurants use MSG liberally. This has resulted in rather widespread concern about this substance among consumers. However, in the most recent review of all 415 GRAS materials, completed in December of 1980, no concerns were expressed nor were modifications made in the GRAS status of MSG, although it is in class 2.

MSG was originally made from wheat gluten but is now made synthetically.

In essence, the flavor enhancers make it possible to use less meat in food formulas, as well as to improve significantly the resulting flavor. Very small quantities of these substances are used, generally under one-quarter of 1 percent.

Chapter 5

DAIRY PRODUCTS AND DAIRY PRODUCT SUBSTITUTES

□ MILK □

Not too many years ago, buying milk was simple, since there really was only one kind. Somebody within a one-hundred-mile radius of our homes milked cows and put the milk in milk cans, which were picked up by a milk truck and taken to a dairy, where the milk was pasteurized and put into glass bottles. Then the milk was brought into the city and delivered to our door by a milkman.

In those days, before homogenization, milk separated into two layers in the bottle: the top layer was cream and the bottom layer was milk without very much cream in it (skimmed milk). We could also get cream delivered to us by the milkman.

Times have changed. The impact of urbanization; fewer, larger farms farther away from population centers; the development of supermarket shopping; the higher cost of everything; diet consciousness; and new nutritional knowledge have tended to produce a more difficult choice for us to make. Homogenized milk, low-fat milk, nonfat milk, light cream, heavy cream, half and half, buttermilk, and a variety of nondairy substitute products are now available. How do they differ? How should they be selected? In this chapter I hope to answer these questions.

■ PROCESSING WHOLE MILK

Homogenization. It all begins with raw milk coming from a real cow. To produce homogenized milk, the raw milk is forced through tiny holes by a very powerful pump. The fat is broken up into such tiny globules that they cannot come back together again to form a layer of fat (cream). No chemicals are added in the homogenizing process. The process is purely physical, and every bit of milk contains the same amount of fat.

Pasteurization. After homogenizing, the milk is pasteurized, that is, heated, in order to kill bacteria that can cause illness, and to reduce the population of all the bacteria.

Milk could be totally sterilized (as it is when making canned milks, such as evaporated milk), but the taste is greatly altered, so we settle for pasteurization as a compromise between best taste and limited shelf life.

Some of us may remember that many years ago milk turned sour in a few days. Now, milk stays sweet for weeks in the refrigerator. There are many reasons for this. Probably the most important is the improvement in cleanliness of milk barns and dairies. Every piece of equipment used for milking cows, as well as transporting, bottling, and handling milk, has been redesigned to make it easy to clean and sterilize.

Also, the pasteurizing process itself has been improved. Nowadays, most milk is pasteurized by heating it to 162° F for 15 seconds, instead of to 142° F for 30 minutes, as was formerly done, with better results in bacteria reduction and taste.

Not least is the improvement in transportation for milk. No longer does the milk sit out in the sun waiting to be picked up by the family: it moves directly from refrigerated dairy into refrigerated cases in the supermarket by way of a refrigerated truck (and *then* into a hot car on the way home).

There is no question that the two steps of pasteurizing milk plus testing cows to make sure they are not carriers of tuberculosis are responsible for wiping out tuberculosis and undulant fever as major diseases in the United States.

The label of homogenized milk shows that something has been added: vitamin D.

Vitamin D. This is known as the sunshine vitamin because it requires sunshine reaching the human skin to produce it in the body. It is not found in many foods, and the foods in which it is found are not terribly popular, such as fish livers.

Insufficient vitamin D during childhood results in rickets: a bone malformation that became more common as people moved into the cities, where they spent more time indoors and got less sunshine.

Because of its importance to growing children, most milk is fortified with vitamin D, and this simple step has virtually eliminated rickets in the United States. The amount used in milk is the recommended daily allowance for children: 400 International Units per quart.

■ LOW-FAT, NONFAT, AND SKIM MILK

To understand the processing of the other forms of milk, first we need to look a bit further into the composition of whole milk. Whole milk, whether homogenized or not, contains all of the butterfat to be found in milk, hence the designation "whole." It also contains proteins and carbohydrates. These three basic substances, fats, proteins, and carbohydrates, are the only significant sources of calories in foodstuffs. (Calories, as we know, are the measure of the fuel value of food. If people eat more calories than they burn off, they will gain weight —and vice versa.)

The fuel value of protein is 4 calories per gram; of fat, 9 calories per gram; and of carbohydrate, 4 calories per gram. Ounce for ounce, carbohydrate is no more fattening than protein! And ounce for ounce, fat is more than twice as fattening as either protein or carbohydrate.

Now, whole milk contains between 3.5 and 4 percent butterfat, which means whole milk is approximately 96 percent fat free. When you see milks advertised or labeled as 99 percent fat free (low-fat milk), that does not mean that 99 percent of the fat has been removed, but that the milk contains 1 percent butterfat instead of the 3.5 or 4 percent.

It's a fact, though, that the fat in milk contributes nearly half

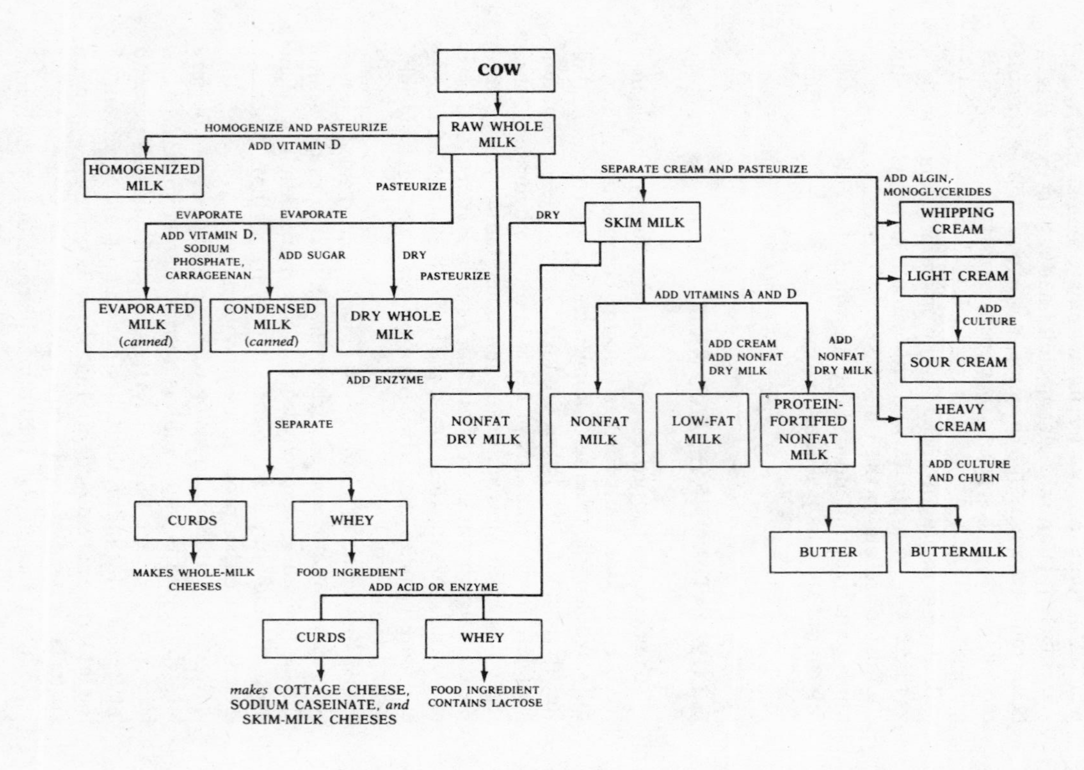
COW
RAW WHOLE MILK
HOMOGENIZE AND PASTEURIZE
ADD VITAMIN D
HOMOGENIZED MILK
PASTEURIZE
EVAPORATE
ADD VITAMIN D, SODIUM PHOSPHATE, CARRAGEENAN
EVAPORATED MILK (canned)
EVAPORATE
ADD SUGAR
CONDENSED MILK (canned)
DRY
PASTEURIZE
DRY WHOLE MILK
ADD ENZYME
SEPARATE
CURDS
MAKES WHOLE-MILK CHEESES
WHEY
FOOD INGREDIENT
SEPARATE CREAM AND PASTEURIZE
SKIM MILK
DRY
NONFAT DRY MILK
ADD VITAMINS A AND D
NONFAT MILK
ADD CREAM
ADD NONFAT DRY MILK
LOW-FAT MILK
ADD NONFAT DRY MILK
PROTEIN-FORTIFIED NONFAT MILK
ADD ALGIN,-MONOGLYCERIDES
WHIPPING CREAM
LIGHT CREAM
ADD CULTURE
SOUR CREAM
HEAVY CREAM
ADD CULTURE AND CHURN
BUTTER
BUTTERMILK
ADD ACID OR ENZYME
CURDS
makes COTTAGE CHEESE, SODIUM CASEINATE, and SKIM-MILK CHEESES
WHEY
FOOD INGREDIENT CONTAINS LACTOSE

of its calories. An eight-ounce cup of milk provides 150 calories. If almost all of the fat is removed, the calories drop to 90, saving 60 calories per cup. To a milk drinker, that can be a big difference.

While there is a tremendous variation in the calorie requirements from one person to another, a good average consumption is 2500 calories per day. (The World Health Organization recommends 2200 for a moderately active woman and 3000 for a moderately active man. The Food and Drug Administration surveys show an average of 2000 calories.) If our average person were a quart-a-day drinker of milk, shifting to nonfat milk would reduce his calories by 240 per day, or almost 10 percent of his total calories! This is enough of a reduction for him to lose weight.

Fat is removed from milk by spinning off the cream in a cream separator. Almost all the fat can be removed this way, so that there is less than one-tenth of 1 percent left. That milk is 99.9 percent fat free and is called nonfat milk or skim milk (the terms are interchangeable). This process, which produces cream as well as nonfat milk, is also purely mechanical. No additives are used up to this point.

Whole milk, however, normally contains vitamin A in the butterfat. This is an important source of vitamin A. When the fat is removed, the vitamin A is removed along with it. For that reason, nonfat milk is fortified with vitamin A, as well as vitamin D, to bring it to the same levels as whole milk.

With growing resistance to the use of animal fats because of worry over cholesterol and saturated-fat consumption, more people tried skim milk but simply did not like it. Nonfat milk, unfortunately, doesn't taste as rich as whole milk and has a watery character that takes some getting used to. Dairy companies came up with more palatable products that overcame many of the objections to nonfat and low-fat milks. For example, instead of removing all of the fat, one-half percent or 1 percent is left in the milk (or added back as cream), and dried nonfat milk is added to improve the taste and consistency.

In addition to low-fat milk, we can also find "protein-fortified skim milk," which is skim milk with nonfat milk solids added.

■ DRY MILKS

Nonfat milk solids are made by drying nonfat milk, usually in a spray tower. The liquid milk is concentrated first by boiling off water under high vacuum and then is sprayed in tiny droplets into a blast of hot air. By the time the droplets reach bottom, they are dry particles and are removed from the dryer. Whole milk is also dried in this manner.

Instant nonfat dry milk may be readily reconstituted by mixing it with water. It is fortified with vitamins A and D and is 25 percent cheaper quart for quart than fresh nonfat milk. The shelf life of dry milk is many months; it can be prepared when needed, and thus, waste may be avoided.

■ SOUR MILKS

Buttermilk. As we know, fluid milks must be kept under refrigeration to prevent their turning sour. But there is one sour milk that we do drink — some of us, anyway — and that is buttermilk. Buttermilk has the same fat content as nonfat milk! It got its name not because it contains butterfat, which it does not, but because it is left over when cream is churned to produce butter. A more apt name would be butter*less* milk.

Buttermilk is now made from pasteurized skim milk by adding a special mixture of bacteria, called a culture, and allowing the bacteria to act on the milk until it curdles and sours. Then it is mixed thoroughly to break up the curds, and packaged.

People generally aren't neutral about buttermilk: there are many who despise it and others who love it.

Buttermilk is not fortified with vitamins, since not very much is consumed and it is not depended upon as a major food source.

It does make a satisfying low-calorie beverage — for those who like it!

Yogurt has been used for centuries and is made by fermenting milk with a mixture of bacteria including Lactobacillus bulgaricus (obviously named after Bulgaria). The milk is usually partially skimmed and sometimes has nonfat milk solids added to build up body. Some brands of yogurt are made from whole milk and this is so stated on the label.

A cup of yogurt made from partially skimmed milk with milk solids added provides 150 calories. As soon as fruit is included, however, the story changes: a cup of fruit yogurt provides 250 to 260 calories.

There is enormous variation in ingredients between brands. One label states: "Peach preserves (peaches), sugar, corn sweetener, pectin, lemon juice." Another, for date-and-nut yogurt, states: "Sugar, nonfat dry milk, modified food starch, gelatin, citric acid, sorbic acid." While there is certainly no reason to believe that there is anything harmful in the second, one must wonder why there is such a difference.

We can deduce that the manufacturers of the second product threw in a little sorbic acid as a preservative to lengthen the shelf life; used citric acid (which is the natural acid in lemon juice) too because they wanted a sour taste without the lemon taste (and it's a lot cheaper than lemon juice); and contributed the gelatin and modified food starch to regulate the consistency and to prevent fluid separation as the yogurt aged.

□ NUTRITIONAL COMPARISON OF MILK PRODUCTS □

One Cup (8 Ounces)

	Fat (*grams*)	Protein (*grams*)	Carbohydrates (*grams*)	Calories	Vitamins Added
Whole Milk	8	8	11	150	D
Nonfat (Skim) Milk	under 1	8	11	90	A & D
Low-Fat Milk (1%)	2	10	14	120	A & D
Low-Fat Milk (½%)	1	9	12	90	A & D
Buttermilk	1	8	11	90	None
Yogurt	4	12	17	150	None

☐ CREAM ☐
Heavy, Light, Whipping, Half and Half, and Sour

The butterfat that is removed with a cream separator when producing skim or nonfat milk (remember, they are the same thing) is used for many purposes: ice cream (which will be discussed in a later chapter), confectionary products, light, heavy, or whipping cream, sour cream, and, of course, butter.

Light cream contains 17 percent fat; heavy, 30 to 38 percent fat; and sour cream, 18 percent fat. Two *tablespoonfuls* of heavy cream have the same number of calories as a full cup of nonfat milk (90 calories). Half and half gets its name from being half light cream and half milk. Two tablespoonfuls contain 36 calories. (Half and half selling at fifty-two cents a pint was found to be eight cents cheaper per pint than buying whole milk and light cream and mixing them together.)

There is a product called "heavy whipping cream — ultrapasteurized." The label shows algin and monoglycerides present. There is also a product called simply "whipping cream"; it is not ultrapasteurized, and it contains nothing but cream. Why are both of these available? Why do we need ultrapasteurization? Why are algin and monoglycerides added? The cost of distributing perishable products from a farm to a big-city supermarket is enormous. Dairy companies are continually searching for ways to reduce the cost of distribution, and one way is to provide longer life for products, making less-frequent deliveries possible. The ultrapasteurized product can be sold for more than six weeks after packaging; the regular pasteurized product for only two weeks. Another reason may be to compete with the nondairy cream substitutes that have a longer life than the dairy products they replace.

Ultrapasteurization, or ultra-high-temperature sterilization, consists of heating the product to more than 200° F for 3 seconds or less. This, combined with extremely sanitary filling procedures, results in a nearly sterile product — with a nearly cooked taste.

In order for the cream to withstand the heat treatment and

retain its natural emulsion, and to prevent it from separating over its now long shelf life, the algin (a stabilizer) and the monoglycerides (emulsifier) are added. Algin is a seaweed extract and is GRAS. Monoglycerides are made from fat and are also GRAS.

In summary, in order to get an extra few weeks of shelf life, the cream is heated to a higher temperature and substances are added to keep the product looking right. Our only loss is the fresh, uncooked flavor!

Sour cream is made by adding a culture to homogenized and pasteurized light cream. The culture is not unlike the one used to make buttermilk. After about 16 hours, the acidity develops that gives the flavor and thickness to the cream, and the product is then chilled and packaged.

□ NUTRITIONAL COMPARISON OF CREAM PRODUCTS □

Two Tablespoons (30 Grams)

	FAT *(grams)*	PROTEIN *(grams)*	CARBOHYDRATES *(grams)*	CALORIES
Light Cream	5	1	1	55
Heavy Cream	9	1	1	90
Half and Half	4	0	2	40
Whipping Cream	10	1	1	100
Sour Cream	5	1	1	55

□ CANNED MILKS □

Some milks are not found in the dairy case, as every mother and many a formula-making father knows. *Evaporated milk* and *condensed milk* are canned and are extremely useful, especially evaporated milk.

The label of evaporated milk says: "Milk, disodium phosphate, carrageenan, vitamin D_3." This product is made by boiling off much of the water in whole milk under high vacuum and then canning. It is twice the "strength" of whole milk, that is, when diluted with an equal amount of water, it has the same solids content as whole milk.

In canning, the milk is heated to a high temperature to make it sterile.

Since milk itself (fresh milk, that is) varies slightly in composition, especially with respect to acidity, it is necessary to control the acidity of the evaporated milk in order to obtain the proper sterilization and stability of the product. Disodium phosphate is used for that purpose, as is sodium citrate in some brands. Both are GRAS. In evaporated milk, for which there is a government standard, not more than 0.1 percent of either sodium phosphate or sodium citrate or a combination of the two is allowed, even though both substances are normal body chemicals.

Since evaporated milk may be stored for a long time, there is a tendency for the whey to separate unless the cans are turned over periodically — a costly procedure. Carrageenan is a seaweed extract that is added to keep the milk solids in suspension and to prevent whey from separating without turning. It is a stabilizer and thickener and is an additive. It once was GRAS.

Condensed milk contains concentrated whole milk and sugar, and is used in cooking, confections, and as a coffee lightener. It contains approximately 44 percent sugar.

□ COFFEE LIGHTENERS □ (Nondairy Creamers)

Because dairy products such as cream and milk turn sour quickly at room temperature, and because there has been publicity advising against the use of animal fats, which butterfat is, a market was created for products that did not spoil rapidly and that contained no animal fat.

Ironically, a major reason for avoiding butterfat is that it is saturated, but some of the lighteners are made with saturated vegetable fat (although the vegetable fat contains no cholesterol).

Coffee lighteners come both wet and dry, and with a variety of fanciful names to conjure up the image of cream, even if they are nondairy.

The ingredient statements of these products show the following materials: water, corn syrup, partially hydrogenated soybean oil (also called "slightly saturated" soybean oil), hydrogenated coconut oil, palm-kernel oil, mono- and diglycerides, soy protein, sodium caseinate, sodium stearoyl-2-lactylate, polysorbate 60, dipotassium phosphate, disodium phosphate, sodium acid pyrophosphate, artificial flavor, artificial colors, beta carotene, sodium tripolyphosphate, sodium silicoaluminate. No one product contains all of the above, but one or two come close.

The price of the liquid products (sold frozen) is less than half that of light cream and slightly more than milk. The dry products cost about the same as the liquid per portion, that is, per cup of coffee lightened.

The only ingredient in these products that occurs as it does in nature is water. All the other ingredients are designed to help provide a colored and flavored emulsion of fat in water to lighten the coffee.

The other ingredients contained are the oils, soy protein, corn syrup, and the sodium caseinate, which are all true foods rather than added substances. Casein is the protein of milk, and is often separated from milk and combined with sodium so that it disperses well, and is used as a source of milk protein in other foods. Mono- and diglycerides have been mentioned earlier — they emulsify; that is, they prevent oil separation. All of these additives are GRAS. Sodium stearoyl-2-lactylate, a material that aids whipping, is a food additive permitted within prescribed limits. Dipotassium phosphate is a sequestrant and emulsifier and is GRAS, as are disodium phosphate, sodium acid pyrophosphate, and sodium tripolyphosphate. Sodium silicoaluminate is an additive permitted at specific levels — all below 5 percent — in some products, and is an anticaking agent. It is used in salt and sometimes in baking powder. The artificial flavors are not identified, nor is the artificial color. The color probably is titanium dioxide — the same white pigment used in paint — and is permitted as a safety-tested food additive at not more than the 1-percent level. While the material is safe, the thought of eating it may be repulsive to many people. Beta carotene is a natural material — the coloring responsible for carrot color — which produces a "brighter" color when used in minute quantities.

Have you tried your coffee black?

□ BUTTER □

Butter is made by adding a bacteria culture to cream to produce the typical flavor of butter and then churning the cream to separate the butterfat from the buttermilk. That butter can then be formed into blocks either with or without salt to produce either salt or sweet butter. The quality of butter is graded according to the Butter Control Act of 1923, which is part of our food laws. Higher score butter is not more nutritious than lower score butter, it is just judged by experts to have better flavor.

Salt is used in butter as a preservative, to lengthen the shelf life of the butter, and for flavor. Butter consists of at least 80 percent butterfat; the balance is the buttermilk that was left in it for appearance, flavor, and consistency. If all the buttermilk were removed, leaving only the butterfat, the product would be called butter oil. Butterfat is an animal fat and is saturated; because of the association with heart disease, its consumption has been dropping. Butter consumption per capita in the United States is only 4.7 pounds per year — less than a fourth that of Denmark, France, and England, one third that of Sweden, and in fact one of the lowest among industrialized countries in the world.

■ DAIRY FOODS AND THEIR RELATIONSHIP TO HEART DISEASE

Milk and butter are major sources of saturated animal fats, as are the high-fat cheeses. The publicity about the relationship of saturated fat and cholesterol (which is also supplied by milk fat) has resulted in important changes in our fat-eating habits, as mentioned in the above paragraph. Now, therefore, seems an appropriate time to discuss this subject in more detail.

Heart disease, stroke, hypertension, and arteriosclerosis are the leading causes of death in the United States. In 1973, over 50 percent of *all* deaths were caused by these illnesses. (I must point out, however, that the life expectancy of a child born in that year in the United States was 71.3 years, which is slightly lower than in twelve other countries, and higher than all the other countries in the world for which figures are reported. The highest life expectancy is reported in Sweden: 74.5 years.)

A heart attack occurs most commonly when a blood clot forms suddenly and blocks a coronary artery already narrowed by accumulated debris at the spot. That debris contains various substances, including cholesterol, fatty acids, compounds of fats and proteins, and other materials. (If the blood clot lodges in a narrowed artery in the brain, the result is a stroke.)

After more than twenty years of research, it is generally agreed that more than one factor is responsible for these diseases. In fact, sixteen different factors have been identified, although there is no agreement among researchers that they are all responsible. There does seem to be general agreement that an increase in blood cholesterol; excessive cigarette smoking; blood pressures above normal; for some age groups, obesity; and family history are the major risk factors. We will confine our discussion of the causes to cholesterol and saturated fats, since they come from foods.

Shortly after World War II, a scientist named Ancel Keys and his coworkers studied the incidence of heart disease related to food intake in a number of countries and concluded that it seemed to be related to the amount of fat in the diet. For example, at the time, the Japanese obtained only 10 percent of their calories from fat, and in the United States we were getting 40 percent of our calories from fat. The Japanese death rate from heart disease was *less than one-seventh* of ours: under one death in a thousand compared with seven deaths per thousand for the United States. It was also learned that the cholesterol levels were lower in Japanese blood serum, and that polyunsaturated fats were a significant part of the fat in their diet. Later research showed the following:

Saturated fats in the diet tend to increase cholesterol.
Polyunsaturated fats tend to reduce cholesterol.
Reduced calories and weight reduction help to reduce cholesterol.
Lower cholesterol intake reduces cholesterol, especially if half the fat in the diet is polyunsaturated.

And in at least one research study, done over a five-year period on over eight hundred men, those who were on high-polyunsaturate diets had lower cholesterol and had fewer deaths from *new* heart disease.

Although it has not been established that a heart attack can be prevented, or that a heart attack would be caused by an increase in cholesterol level, the evidence indicates that it is smart to keep down the amount of fat, cholesterol, and saturated fat in our diets. Therefore, the American Heart Association recommends the following preventative daily dietary pattern:

Class of Food	Amount in Grams	Calories
Carbohydrate	273	1092
Protein	114	456
Fat	83	747

Total calories per day: 2295
Percentage of calories from fat: 33
Saturated fat: 7
Polyunsaturated: 10
Mono- and diunsaturated: 16
Ratio of polyunsaturated to saturated: 1.4 to 1

The question now is: How in heaven's name can we tell *what* fats we *are* eating, and whether they are the right ones? To answer this question, we need to understand a little more about saturated and unsaturated fats.

Cholesterol levels in the blood are determined by a number of factors: the amount of cholesterol actually eaten in foods, the total amount of fat we eat, the ratio of saturated to unsaturated fats, and, apparently, the amount of exercise we get.

The saturated fats are those to which no hydrogen can be added. Monounsaturated fats are those to which hydrogen can be added at only one place in the molecule. Polyunsaturated fats are those to which hydrogen can be added at more than one point in the molecule.

The saturated fats raise cholesterol levels, the polyunsaturates lower them, and the monounsaturates seem to have no effect. The fats we find in foods generally are mixtures of all three.

Butterfat — whether in butter or milk — contains 50 percent saturated fat, about 5 percent polyunsaturated fat, and about 45 percent monounsaturated fat. This gives us a ratio of polyunsaturates to saturates of 1 to 10, rather than the recommended 1.4 to 1. In addition, butter supplies 35 milligrams of cholesterol in a tablespoonful: about 10 percent of the recommended maximum daily intake.

Labels for foods with high fat content disclose the amounts of polyunsaturates and saturates, or the ratio of polyunsaturates to saturates, as well as the amount of cholesterol in the food. The labels can help you to choose your foods to try to meet the recommended levels of fat and cholesterol intake.

The growing awareness of the significance of fat and cholesterol in the diet has greatly increased the demand for margarine.

□ MARGARINE □

For many years margarine could not be sold already colored. It was sold as a white, opaque substance, together with a small capsule of coloring material. The consumer had to burst the capsule and mix the white margarine with the coloring material. This particular requirement was the result of pressure by the dairy lobby in a number of states, but progress cannot be stopped, and now we can buy completely artificially colored margarine!

Here is a typical margarine ingredient statement: "Partially hydrogenated soy oil, water, salt, nonfat dry milk, lecithin, mono- and diglycerides, sodium benzoate, citric acid, artificial flavor and color, vitamin A palmitate." The label also states "no cholesterol." Other oils used in margarine are corn oil and sunflower oil, and other ingredients that appear on some labels are isopropyl citrate, calcium disodium EDTA (to protect flavor), potassium caseinate, and whey. All margarine labels state the amount of saturated and unsaturated fat in a tablespoonful (14 grams). Margarine, like butter, contains 11 grams of fat per tablespoonful. And, margarine, like butter, provides 100 calories for each tablespoonful (enough to spread four slices of bread).

There is soft margarine as well as hard margarine. The soft margarine has slightly more polyunsaturated fat than the hard. For example, one manufacturer's soft margarine has 5 grams of polyunsaturated in a tablespoon and the same brand of hard margarine has 4 grams of polyunsaturated fat.

To make a product that resembles butter in flavor and spreading characteristics and that doesn't spatter when the product is used for pan frying, a number of additives are required. Surely, if margarine were sold only as a white, tasteless material that spattered when you fried an egg in it, very few people would buy it or use it, regardless of its real or purported health advantages over butter.

Let's go through the additives to learn more about what they do in margarine. Water is added to margarine in order to give it the same consistency as butter. It's used to help adjust the fat content to exactly the same percentage as that of butter. Salt is added for flavor as well as to improve shelf life slightly. Nonfat dry milk is added to impart the milk taste that is present in butter. The nonfat dry milk in the margarine together with the water should roughly approximate the buttermilk content of butter. Lecithin is added to improve the spreading characteristics of the margarine. (Lecithin is actually a food rather than an additive, although it doesn't appear in the pure state in nature. It is extracted from soybean oil and is present in many unrefined oils and fats.) Mono- and diglycerides are emulsifiers and help to prevent spattering of the margarine. Sodium benzoate is present to prevent yeast and mold growth and to lengthen the shelf life; and citric acid is there to help adjust the acidity of the product, primarily for flavor. Artificial flavor and color are added to increase the resemblance to butter. The vitamin A palmitate is added to duplicate the level of vitamin A found in butter. Isopropyl citrate is a derivative of citric acid and prevents oxidation. Calcium disodium EDTA is a metal sequestrant and prevents the destruction of flavor by any metal contaminants in the fat. It also increases flavor stability. Potassium caseinate is a milk-protein derivative and is sometimes used, together with whey, instead of nonfat dry milk. Why a substitute for nonfat dry milk? Simple answer: potassium caseinate and whey cost less than nonfat dry milk—but do not conclude that the product containing whey and potassium caseinate will sell for a lower price!

What possible conclusions can we draw from this information about choosing margarine or butter? Well, as far as the Food and Drug Administration knows, all of the additives in margarine are safe at the levels at which they are present. I don't think anybody knows or can know whether these additives are harmful or harmless in any individual diet when they are combined with other foods that may contain other additives, but there certainly is no evidence of any kind that any of these additives are harmful in themselves. The nutritional value of margarine and butter are identical.

The differences lie in cholesterol, saturated and polyunsaturated fats, and flavor. In order to decide which product to use, you must know yourself. Do you eat lots of eggs? If you do, then consuming more cholesterol in butter is probably undesirable. If you don't eat many eggs and don't get very much fat from other foods, butter should be fine. We can't make a flat statement that "butter is bad." If every other factor identified with stroke and heart disease is absent in your case (for example, if you don't smoke, if you get plenty of exercise, are not under great stress, and are not overweight), eating butter should cause no problems. But if you are overweight and do smoke, or are otherwise already at risk, then it would be wise to restrict your intake of saturated fats, and use margarine instead of butter.

□ SUBSTANCES ADDED TO MILK AND MILK PRODUCTS □

	MATERIAL ADDED	LEGAL STATUS (*and GRAS class*)	PURPOSE
Whole Milk	Vitamin D_3	GRAS (2)	Nutrition
	Vitamin D_2	GRAS (2)	Nutrition
Low-fat Milk	Vitamin A palmitate	GRAS (2)	Nutrition
	Vitamin D_3	GRAS (2)	Nutrition
Nonfat or Skim Milk	Vitamin A palmitate	GRAS (2)	Nutrition
Evaporated Milk	Algin	GRAS (2)	Thickener
	Carrageenan	Additive (3)	Thickener
	Monoglycerides	GRAS (1)	Emulsifiers
	Sodium citrate	GRAS (1)	Acidity control
	Sodium phosphate	GRAS (1)	Consistency regulation
Yogurt (with fruit)	Citric acid	GRAS (1)	Acidifier (makes tart)
	Modified food starch	Additive (1–5)*	Thickener
	Pectin	GRAS (1)	Sets fruit jelly
	Sorbic acid	GRAS (1)	Preservative
Coffee Lighteners	Beta carotene	GRAS (1)	Coloring
	Dipotassium phosphate	GRAS (1)	Acidity control
	Disodium phosphate	GRAS (1)	Acidity control
	Polysorbate 60	Additive	Emulsifier
	Sodium caseinate	GRAS (1)	Flavor and color
	Sodium pyrophosphate	GRAS (1)	Sequestrant
	Sodium silico-aluminate	GRAS (1)	Free-flowing agent
	Sodium stearoyl-2-lactylate	Additive	Emulsifier, whipping agent

*Modified food starches are not now specifically identified on the label. They may be anything from class 1 to 5, depending on which starch is used.

Chapter 6

CHEESES

Cheese is one of our most important foods. It provides a source of high-quality protein as well as a tremendous variety of pleasing tastes. Man has used cheese for well over 4000 years. (In the case of some types of cheese, it's one way to preserve milk in a warm climate.)

Dairy foods are the principal source of calcium in the diet, and one of the four basic food groups we should choose from every day. But many people have difficulty drinking milk, especially in their later years. Cheese, particularly American-style cheese, can be a crucially important source of calcium to these folks. One slice of regular American cheese supplies 15 percent of the RDA of calcium. A slice of the new low-fat American-style cheese supplies between 10 and 13 percent.

■ CURDS AND WHEY

It is the protein of milk, casein, that makes cheese possible. When Little Miss Muffet sat on a tuffet (whatever that is), eating her curds and whey, luckily she was not aware of the fact that she was *really* eating her coagulated casein (curds), and lactose and milk-albumin solution (whey).

When an acid or an acid-producing bacterium or a clotting

enzyme like rennet (which is extracted from the stomach lining of sheep) is added to milk, a curd is formed. The free fluid that is left, plus that which is drained from the curds, is whey.

If the curd is made from whole milk, the butterfat of the milk is trapped and included in the curd. If, however, the milk is skimmed before curding, the curd is fat free.

So we have whole-milk cheeses, skim-milk cheeses, and partly skimmed milk cheeses — each with a different fat content.

■ SOFTNESS

Cheeses also differ in their softness or hardness, which is the result of how much whey is removed and how dry the cheese is permitted to become; very simply, the softness or hardness depends on the cheese's water content.

Cheeses also differ in spreadability, which is usually affected by the ripening of the cheese with age, its hardness, and the exact bacterial culture used to make it.

■ CHEESES ARE STANDARDIZED FOODS

Cheeses, under our food and drug laws, are standardized foods. This means there are published standards of identity for them that must be met in order for the food to be called cheese of any specific type.

The standards of identity include methods of manufacture, ingredients that are required, ingredients that are optional, and the precise labeling requirements, from the product name to how each of the ingredients must be declared.

One of the oddities of our food laws is that our labels tell us less about the composition, nutritional value, and ingredients of our unprocessed, unpackaged foods and of our standardized, packaged foods than about processed, packaged, unstandardized products. For example, it is *not* now necessary to declare the use of artificial color in a number of the standardized cheeses! For foods that are subject to standards of identity, it is *not* all on the label; and for unpackaged foods, there is no label.

Of course, anybody who wants to buy a copy of the Code of Federal Regulations, Number 21 (Food and Drugs), can look through the more than one thousand pages and read the standards of identity and learn far more than any consumer would want to know. For example, in defining milk for mozzarella cheese:

> Milk shall be deemed to have been pasteurized if it has been held at a temperature of not less than 145° F for a period of not less than 30 minutes, or for a time and temperature equivalent thereto in phosphatase destruction. The finished food shall be deemed not to have been made from pasteurized milk if 0.25 gram shows a phenol equivalent of more than 3 micrograms when tested by the method prescribed in P 19.500 (f) provolone modification.

The original idea in setting up standards of identity was to protect the consumer. Although the standards are meant to protect us, in meeting the standards, products did tend to become alike, and information that we may want to have is unavailable to us on the label.

But on to the cheeses.

□ THE UNCURED CHEESES □

Curing means storing the cheese to give the bacterial culture that makes the cheese unique a chance to work. Some very popular cheeses are uncured. They are made and sold as fresh as possible. The most common is cottage cheese.

■ COTTAGE CHEESE

Cottage cheese begins with pasteurized skimmed milk. One method calls for lactic-acid-producing bacteria to be added, with or without rennet enzyme, which aids in curd formation.

This forms a curd, which is drained, cut, pressed, chilled, and seasoned with salt.

Another way is to add acids, with or without rennet, to the pasteurized skimmed milk. No bacteria are used. The acid is not the lactic acid formed by the bacteria in milk, but phosphoric, citric, or hydrochloric acid, or synthetic lactic acid. When this procedure is used, the cheese label must say "directly set" or "curd set by direct acidification."

Either method produces dry-curd cottage cheese. This must contain less than 0.5 percent fat and not more than 80 percent water by law.

Table-ready cottage cheese is made by mixing the dry-curd cottage cheese with a creaming mixture of milk and/or cream and/or dry skimmed milk and water, plus other ingredients to be described a bit later.

Cottage cheese contains 4 percent fat — the same as whole milk — and not more than 80 percent water. (Whole milk contains 87 percent water.)

Low-fat cottage cheese contains from 0.5 to 2 percent fat, and 82.5 percent water.

Let's compare some labels:

Brand A Cottage Cheese, 4% fat: Skim milk, milk, cream, salt, xanthan gum, enzyme.

Brand A Cottage Cheese, 1% fat: Skim milk, milk, cream, salt, xanthan gum, enzyme, artificial flavor.

Brand B Cottage Cheese, 4% fat: Cultured pasteurized skim milk, pasteurized milk, cream, salt.

All are cottage cheese. None is dangerous to your health. Nutritionally, all are magnificent foods. For well under a dollar (at the time of publication), you can buy the recommended daily allowance for protein of top quality, without the undesirable load of fat in meat, in a form that mixes well with many other types of foods to produce healthy, balanced meals: with pasta, fruits, green vegetables.

Which brand would you rather eat, and why? Unless there were a major price difference or flavor improvement in Brand

A, which contains xanthan gum, I'd pick Brand B — it's simpler. The only reason Brand A adds the xanthan gum is to delay weeping of the cottage cheese (the separation of the fluid — whey — from the cheese) under prolonged storage. Xanthan gum is not GRAS, but is approved as a food additive.

But why buy a product with that additive when another product that tastes as good, costs the same, and doesn't contain the additive is available?

What about the difference between the 4-percent-fat cottage cheese and the low fat? A portion of cottage cheese is usually half a cup, weighing about 115 grams, or 4 ounces. Let's compare the calories. One-percent-fat cottage cheese contains 3.3 grams less of fat in a portion and saves 30 calories. While this is nearly 30 percent of the calories in a portion, still it is only 30 calories — less than 1.5 percent of the day's calorie needs.

Since *any* cottage cheese is already a major calorie saver compared to other high-protein foods such as meat or high-fat cheeses, my choice would be the cottage cheese I like best — unless I were very strictly and desperately counting every last calorie.

■ CREAM CHEESE

Cream cheese is an uncured cheese that is made by blending milk and cream so that the finished cheese ends up with 33 percent fat. Whereas cottage cheese represents the low-fat end of the uncured cheese group, cream cheese is the highest-fat uncured cheese, containing more than seven times the percentage of fat of cottage cheese and four times the number of calories.

Cream cheese labels will read: "Cultured milk and cream, salt, enzymes, and carob-bean gum." The enzyme is probably rennet, and the carob-bean gum is produced from carob-bean seed and is GRAS. The pod containing the seeds is known as Saint-John's-bread.

The purpose of the gum is to control the texture of the cheese so that it spreads smoothly. Old-fashioned cream cheese tends to be a bit crumbly. Some folks think that the presence of gum prevents the full release of the cheese flavor — that is, it destroys or damages the "melt in your mouth" effect.

There is a spreadable cheese that lies between cottage cheese and cream cheese, containing less fat than cream cheese: Neufchatel. This cheese contains 80 calories per ounce compared to 100 for cream cheese. The ingredient statement for Neufchatel is likely to be identical to that of cream cheese; it simply has less cream and more milk.

You may also buy imitation cream cheese, which has half the calories of real cream cheese. The ingredient statement on one brand reads: "Cream cheese, cottage cheese, water, skim milk, lactic acid, sodium citrate, salt, carob-bean gum, artificial flavoring." Imitation cream cheese appears to be diluted with water and skim milk, and then flavored and stabilized to imitate the flavor and texture of the standard cream cheese. All the ingredients are GRAS. The lactic acid and sodium citrate are added to give the product the same acidity as cream cheese, and the artificial flavor to compensate for the flavor diluted by the water.

Another variation on the theme is whipped cream cheese, labeled "Cream cheese (pasteurized milk and cream, cheese culture, salt, carob-bean gum), skim milk, salt, guar gum." There was no nutritional analysis on the package, so we don't know how much the cream cheese has been diluted with skim milk. The guar gum is added as a stabilizer to keep the thing whipped; that is, to hold the air that has been whipped in. Guar gum is extracted from the seed of the guar plant, widely used as a cattle food in India and Pakistan. It is GRAS.

For those who watch their calories, substituting any of the above cheeses for butter or margarine represents a major calorie reduction.

Spread	Calories per Ounce
Butter	200
Cream Cheese	100
Neufchatel	80
Imitation Cream Cheese	50

None of the above cheese is a significant source of other nutrients in the quantities normally eaten, so that it is probably a good idea to pick your uncured-cheese spread based on taste and calories, and price. On the day I checked the imitation cream cheese versus the regular, the price was *higher* than the regular and the Neufchatel! The whipped cream cheese sold for 20 percent more per ounce than the regular cream cheese.

■ MOZZARELLA

With the pizza explosion in the United States, mozzarella has become an important cheese to us, not only in pizza, but in hot dishes at home. (Try refrigerated mozzarella sprinkled with a bit of salt and pepper as an hors d'oeuvre — it's great!)

Mozzarella is also an uncured cheese. The texture comes from mixing the curd to make it somewhat rubbery; the kneading action alone does it.

This cheese is made from whole milk, or partly skimmed milk, and is either high or low moisture. The moisture content depends on how much whey is separated from the curd. If the curd is squeezed pretty dry, it will be low moisture.

While none of the packages viewed showed the nutritional analysis, the government standards for the different styles of mozzarella indicate the following analysis:

TYPE	% FAT	% WATER	CALORIES PER OUNCE
High Moisture/Whole	18	60	71
High Moisture/Part Skim	12	60	63
Low Moisture/Whole	21	52	85
Low Moisture/Part Skim	17	50	80

If a person ate as much as 4 ounces of mozzarella, which is a lot, the maximum calorie difference between the lowest-calorie and highest-calorie product would be 88 calories, which is significant. It might be simpler to remember that the part-skim-milk mozzarella is the lowest-calorie product.

The ingredients of mozzarella are not always on the label because they are not required, but some brands state them anyway: "Pasteurized cultured milk, salt and enzymes, calcium chloride."

Sorbic acid or potassium sorbate is permitted in mozzarella as a preservative to prevent mold growth, and when present, it must be declared.

□ UNCURED CHEESE ANALYSIS □

PRODUCT	% FAT	% WATER	CALORIES/ OUNCE	CALORIES/ SERVING	SERVING SIZE *(ounces)*
Cottage Cheese	2	82	25	100	4
Creamed Cottage Cheese	4	80	30	120	4
Cream Cheese	33	55	100	100	1
Neufchatel	20	65	80	80	1
Mozzarella					
High Moisture/ Whole	18	60	71	71	1
High Moisture/ Part Skim	12	60	63	63	1
Low Moisture/ Whole	21	52	85	85	1
Low Moisture/ Part Skim	17	50	80	80	1
Ricotta	11	70	48	190	4

□ THE CURED CHEESES □

These are the cheeses with a strong cheese taste, from the mildest cheddar to the most potent Roquefort. The flavor of each cheese is produced by a specific strain of bacteria or mold. The cured cheeses are generally made in the following manner:

1. Milk is pasteurized, and sometimes bleached. If the milk is not bleached, it can be artificially colored.

2. The cheese culture (containing the right kind of bacteria or mold), lactic-acid-forming bacteria, and, in some cases, propionic-acid-producing bacteria, are added to the milk.
3. Rennet, or some other clotting enzyme, is added.
4. The curd is separated, with or without pressing, with or without salting, with or without mixing. (These steps vary from one type of cheese to another.)
5. The curd is formed into the desired cheese shape.
6. After it is formed, it is stored under temperature and humidity conditions that are favorable to the bacteria or mold being used to produce the specific cheese.
7. The cheese is then aged. (Such aging would simply spoil uncured cheeses.) In some cases, the salted curd is stored for a number of days first.

The following ingredients may be added and not declared: calcium chloride, artificial color, potassium alum, calcium sulfate, plus cheese-coating materials: magnesium carbonate paraffin, or vegetable fat, or color in the coating. The exact bacteria (or mold, in the case of blue cheese and Roquefort) used is not declared.

Among the ingredients that must be declared are sorbic acid, and benzoyl peroxide used for bleaching the milk. Since bleaching destroys the vitamin A originally present in the milk, it is necessary to add vitamin A to bleached milk.

The necessity of bleaching milk is something I find difficult to accept, and although the amount of benzoyl peroxide used is minute (not more than 2 parts per 100,000), and there is no evidence of danger, this is not the same as proof of safety. Furthermore, the "benzoyl" part of the molecule is not a normal body substance, the compound has not yet been reviewed, and I see no point in being exposed to the substance, along with possible human error in its use, when there is no benefit other than the questionable one of color.

If sorbic acid is used, as is permitted for sliced cheese, not more than 3 parts in 1000 are allowed and must be declared. (When cheese is sliced, the chances of mold contamination are greatly increased, since there are more exposed surfaces. Sorbic acid acts as a mold inhibitor.)

Calcium chloride, used to firm up the curd, is limited to less than 2 parts in 10,000. Although the quantities used are not sufficient to supply nutritional calcium in significant amounts, the substance is absorbed and used for bone building.

■ AMERICAN CHEESE

This is the most popular cheese in the United States. Most of us recognize it as the white or yellow square slice that used to be purchased in a two-pound block in a box (still available in some stores). Today, what most of us call American cheese is usually offered already sliced, and even with slices individually wrapped — an investment of more of our energy and money in packaging materials, most of which are made from imported oil! Some people prefer white American cheese, others yellow, not realizing that the only difference is the added color.

It is getting more and more difficult to make sense out of the variety of those square slices that we call American cheese. As the chart below shows, there are quite a few American cheese products. Three forces have come about to cause this proliferation of American-cheese-like products. First is the ever-growing pressure to get or keep the cost of cheese down. Second is the growing market for low-fat foods, especially foods low in dairy fats. (Just recently, an imitation cheese made with corn oil has been introduced.) The third force is the effort to fill the various market needs for convenience: in spreading, avoiding slicing at home, preventing slices from sticking together.

It's interesting to compare the nutritional analyses of the American-type cheeses. I have included the analyses for colby, cheddar, and Monterey Jack, since those are the true "pure" cheeses, not processed, and are the cheese ingredients of the "pasteurized process" products.

Pasteurized process American cheese is made by adding a small amount of water to cheddar or colby or Monterey Jack. The cheese foods and cheese products are so called because they do not meet the government standards for cheese, especially with respect to fat and water content. Diluting the

□ AMERICAN CHEESES □

	Calories/ Serving	Fat/Ounce (grams)	Protein/Ounce (grams)
Monterey Jack	100	8	6
Colby	110	9	7
Cheddar	110	9	7
Process American Cheese	110	9	6
Process American Cheese Food	90	7	6
Process American Cheese Spread	80	6	5
Process American Cheese Product	50	2	7
Skim American Cheese Product	70	5	6

cheese ingredient further produces American cheese "food"; still greater dilution yields the cheese "product"; and cheese "spread" has the lowest percentage of pure cheese. As you can see, you can buy a food that looks like American cheese — process American cheese food — with fewer calories and lower fat, that still contains most of the high-quality protein. The reason for the difference in calories between the pasteurized process American cheese and the cheese "food" and "product" is that there is less fat and more water in the diluted products.

It is quite revealing to compare the ingredients of the various cheese products as shown on the labels.

Monterey Jack: Cultured pasteurized milk, salt, enzymes, calcium chloride.

Pasteurized process American cheese: American cheese, water, cream, sodium phosphate, salt. (Another brand, a sliced product, also adds kasal, sorbic acid, and sodium citrate.)

Colby: Pasteurized milk, cheese culture, enzymes, salt, calcium chloride.

Cheese spread: American cheese, whey, water, skim milk, sodium phosphate, cream, salt, oleoresin paprika.

Skim American product: American cheese, skim-milk cheese,

water, sodium citrate, sodium phosphate, salt, sorbic acid, xanthan gum, carob-bean gum, guar gum.

Pasteurized process cheese product (low calorie): Skim-milk cheese, water, American cheese, sodium phosphate, enzyme modified cheese solids, salt, sorbic acid, citric acid. Artificially colored.

The simplest products have only sodium phosphate added. Our bodies need phosphorus, and the only way we get it is through phosphates, found either naturally, as sodium phosphate or calcium phosphate or other forms of phosphate, or through the addition of calcium phosphate. It is used to stabilize the acidity of the cheese and thus helps to assure a uniform product with uniform consistency.

One brand of processed American cheese also contains sorbic acid, sodium citrate, and kasal. This last is a misleading name. It is actually potassium aluminum sulfate — "alum" to you — which, together with hydrogen peroxide, is used to bleach the milk from which cheese is made.

Sodium citrate is used to control acidity and is GRAS. Like so many synthetic additives, it is a normal component of human blood and muscle, and when used in moderation is completely harmless.

When cheese is subjected to fluctuating temperatures — removing from and replacing in the refrigerator — water will condense on the surface of the cheese, and this will promote the growth of molds. The sorbic acid stops or, at any rate, delays this.

As water is added and the cheese becomes cheese food, citric acid is required, because as water is added, the natural acid in the cheese is diluted and needs to be strengthened for flavor. Citric acid is formed in lemons and oranges, and is also made synthetically. It is a normal body component and is GRAS.

Some of the cheese products contain xanthan gum, carob-bean gum, or guar gum. The gums are needed to produce an acceptable texture in these products as the amount of water is increased. (Water itself may be added; or whey, which is 96 percent water, or skim milk, which is 92 percent water, may be used.) Xanthan gum is man-made — or, more accurately, is

□ SUBSTANCES ADDED TO CHEESE □

Material Added	Legal Status (and GRAS class)	Purpose
Calcium chloride	GRAS (1)	Firming agent
Calcium sulfate	GRAS (1)	Firming agent
Carob-bean gum	GRAS (2)	Thickener
Citric acid	GRAS (1)	Acidifier
Guar gum	GRAS (2)	Thickener
Hydrochloric acid	GRAS (1)	Acidifier
Hydrogen peroxide	GRAS (2)	Bleaching agent
Kasal (potassium aluminum sulfate or alum)	GRAS (1)	Hardening agent
Lactic acid	GRAS (1)	Acidifier
Magnesium carbonate	GRAS (1)	Acidity adjustment
Phosphoric acid	GRAS (1)	Acidifier
Potassium alum (same as kasal)	GRAS (1)	Firming agent
Potassium sorbate	GRAS (1)	Preservative
Rennet	GRAS (1)	Causes curding
Salt	GRAS (4)	Seasoning
Sodium citrate	GRAS (1)	Acidity control
Sodium phosphate	GRAS (1)	Consistency regulation
Sorbic acid	GRAS (1)	Preservative
Xanthan gum	Additive	Thickener

made by bacteria under man's guidance. The other gums are extracted from plants. The xanthan is legally an additive; the others are GRAS, in class 2.

■ CALORIES AND FAT IN CHEESE

From a practical point of view, you can ignore the difference in calories among all of the following cheeses: Swiss, Gruyère, Muenster, Edam, Gouda, blue, Roquefort, provolone, Monterey Jack, Parmesan, cheddar, pasteurized process American.

What you should *not* ignore is the amount you eat. Chances are you will eat far more Muenster (because the flavor is mild) than blue cheese — and far more of those two than Parmesan, which is usually grated.

To cut calories among the cured cheeses, go to the cheese foods or the cheese products, or eat less.

Cheese represents a most important source of protein, calcium, and, in the case of the cured cheeses, a most significant source of fat in our diets. The recommended amount of calories we should get from fat is around 35 percent of our total calories. Each time we eat a cured cheese — which is between 24 and 30 percent fat — we actually get 70 percent of the calories in the cheese from fat. We need to eat an awful lot of no-fat foods to get down to the 35 percent level after eating high-fat cheese. Unfortunately, the high-fat cheeses are delicious, and we tend to nibble away at them, especially at cocktail time.

Even cottage cheese with 4 percent fat supplies 33 percent of its calories from fat!

If we are to make use of our new knowledge about labels, and at the same time heed the warnings of the best scientists of our time to keep our fat intake down, we may need to budget our cheese-fat intake.

Chapter 7

BAKED FOODS

Baked foods are available to us from many different sources. We can go to the retail bakeshop and buy unwrapped bread and other products, and since they have no labels, there is no way to tell what the ingredients are. Or, in many supermarkets, there are shops called "in-store bakeries," which sell unwrapped, and sometimes wrapped, products made in the store. They may or may not have labels and ingredient listings. In the freezers, there is a host of baked products, produced by special frozen-baked-food manufacturers, all meeting, or at least required to meet, the labeling requirements of the FDA. And finally, there are the products that are produced fresh in manufacturing bakeries, then packaged, labeled, and distributed to grocery stores and supermarkets.

Do not make the mistake of thinking that the products made in the bakeshop contain different ingredients from those produced by the big bakeries — they merely don't supply the information. The ingredients could be identical, but you have no way to tell. I think it would be a good idea for retail bakeshops and all others that sell unlabeled products to display ingredient listings in the shops. It really ought to be required by law.

Nearly 80 percent of all baked foods that are purchased are the factory-manufactured kind. Bread baking is not easy, and it takes a lot of time. For that reason, the first and most important convenience food was bread baked by someone other than the home baker. The bread manufacturing industry is very old, going back to Roman times. (Names like Miller and Baker come from the occupations of ancestors who helped to produce bread for others.) While there is a good deal of charm and fun in baking at home, it is a fair bet that we will continue to depend on bakeries and large manufacturing plants to supply our baked foods.

Before we look at the ingredients in baked foods, it is helpful to know what the baking process is, since so many of the ingredients used are added for processing and no other purpose.

■ BREAD AND ROLLS

The dough is mixed and allowed to ferment for a few hours until it rises. Then it is divided by machine into pieces of the correct weight for each loaf. The pieces are spread into a sheet and rolled up into a cylinder by a molder and dropped into a baking pan. The baking pan is then placed in a proofer — a warm, humid chamber — and the dough is allowed to rise again, for about an hour. The pan is then fed to an oven set at about 400° F, where the bread is baked for 25 to 35 minutes, depending on the type. After baking, it is removed from the pan and placed on a conveyor belt that takes the bread for a ride through a cooler for an hour or so. The cooled bread is fed to a slicer and wrapper, and loaded on a truck for transport to stores.

Another method of preparing the dough is to feed all the ingredients to a continuous mixer, which mixes and deposits pieces of dough directly into the pan. This eliminates the first fermentation step — and a good deal of the flavor.

■ COFFEE CAKE AND DANISH PASTRY

The dough is mixed and fermented, very much like bread, and then rolled into a sheet through rollers feeding a conveyor belt. Fillings are spread on the sheet, and the sheet is then rolled, folded, or otherwise sealed, depending on the variety or shape. The pieces are then cut by either machine or hand and placed on a pan for proofing. The pans remain in the proofer for one hour to let the dough rise, and then the dough is baked. After baking, icing or topping is applied, and the product is cooled and then packaged.

In the case of Danish pastry, the dough is sheeted three times: butter, margarine, or shortening is spread on each time the dough is folded over, and is placed in the refrigerator between each sheeting. This repeated sheeting and refrigeration is done to prevent the fat from migrating throughout the dough and to make the finished product flaky. After sheeting, the dough is divided and shaped and thereafter treated the same as coffee cake.

■ DOUGHNUTS (Yeast Leavened — The Fluffy, Raised Variety)

These are prepared like rolls, except that the doughnuts are formed by special machines, and after processing, the doughnut is fried in deep fat rather than baked. Glazed doughnuts are dipped before cooling in a mixture containing about 80 percent sugar.

■ ENGLISH MUFFINS

These, too, are treated like rolls, but baked on a griddle instead of in an oven.

■ CAKES, MUFFINS, AND BISCUITS

The batter is mixed, poured into a pan, baked, and cooled, and the finished products are packed.

■ COOKIES

The batter or dough is mixed and then cut or dropped directly onto a band that travels through an oven and discharges the baked cookie in a cooler at the other end. Then a conveyor carries the cookies to be packaged.

■ CRACKERS

The dough is mixed, fermented, sheeted, proofed, scored, baked on a band traveling through an oven, cooled, and packaged.

□ BAKING INGREDIENTS □

The basic ingredients of most baked foods are the same: flour, sugar, shortening, milk, eggs, salt, baking powder or yeast, flavors, spices.

The principal variations between baked foods are in the amounts of each ingredient present, the process used, and whether the product is leavened with baking powder or yeast or not at all.

Later in this chapter, the difference in amounts will be given in detail. At this point, your understanding of labels will be enhanced by learning more about some of the ingredients themselves.

■ YEAST

Bread and breadlike products such as rolls, coffee cake, Danish pastry, yeast doughnuts, and English muffins are each leavened with yeast. Yeast is a living microorganism that feeds on the sugars in dough to convert that sugar to carbon dioxide (a gas), alcohol, acetic acid, and many other products that provide the delicious aroma and taste of yeast-leavened baked foods. The yeast action usually takes hours. The sugar the yeast needs to work on is not normally present in flour but must be produced by enzymes already in the flour, or may be added to the flour, to release sugar as food for yeast. The sugar is produced by the action of enzymes on the starch in the flour.

While it is not catastrophic if there is not enough yeast food naturally present, the timetable for baking could be delayed for hours, so it is common for *yeast nutrients* to be added: calcium sulfate, ammonium sulfate, calcium phosphate. As you can see, calcium is very important to the working of yeast. Naturally hard water has these very minerals in it. Bakers in different parts of the country use different amounts and types of yeast food to fit their water supply.

■ BAKING-POWDER INGREDIENTS

Whether yeast or baking powder is used, the gas that expands the dough is the same: carbon dioxide. When chemical leavening (baking powder) is used, the carbon dioxide almost always comes from common baking soda: sodium bicarbonate (bicarbonate of soda). (Carbon dioxide and bicarbonates are native to our bodies. Each time we inhale, we include some carbon dioxide from the atmosphere and form some bicarbonate in our bloodstream.)

However, sodium bicarbonate needs an acid to release the carbon dioxide from it. Any acid will produce this gas when it is brought into contact with the sodium bicarbonate: vinegar; the citric acid in lemon juice; tartaric acid, which is extracted from grapes; or the lactic acid in buttermilk. Baking powder is a mixture of sodium bicarbonate and one or more of the baking acids, diluted with starch or flour so that it is not too concentrated for home use.

Leavening with baking powder takes seconds rather than minutes or hours as in the case of yeast, but even so, there must be control within those seconds. After a batch of cake batter is mixed for commercial baking, it takes a while for all the cake pans to be filled from that batch — perhaps 20 to 30 minutes. If the gas were formed too fast, it would be gone before the batter at the tail end of the batch got to the oven, and the cakes would not rise properly.

Also, if all the carbon dioxide were formed before the batter was in the oven, the early part of the baking process would simply drive the gas off, leaving none for raising the batter when it begins to thicken and set during baking. So, double-acting baking powders are used, with some gas formed during mixing to produce the bubbles wanted in the batter, and more formed at the desired temperature and time in the oven, so that the thickening batter can be expanded and held that way until it sets into a cake.

This accounts for the addition of materials like calcium acid phosphate (fast acting), slow-acting sodium acid pyrophosphate, slow- and fast-acting sodium aluminum sulfate, and slow-acting glucono delta lactone, which are the baking acids used to react with the sodium bicarbonate at the appointed time in the mixing and baking cycle. Since each of these baking acids reacts under somewhat different conditions of temperature and at different speeds, they are used in different products as needed for lightness and fluffiness.

Clearly, baking is simpler when done in small batches, when one can expect the batter to be in the oven minutes or even seconds after mixing.

■ FLOUR

The treatment of this ingredient has become an emotional issue. How often have you read the accusation that the millers have robbed our flour of its natural goodness, removed all of its nutrients, and put back only a few as enrichment?

White flour has long been preferred to whole-grain flour as soon as it became affordable. As every society became industrialized, its people have wanted white bread.

Upon its introduction, opposition to mass-produced white flour was expressed in two countries — the United States and England — but not in other countries. The principal argument, then as now, was that whole-wheat flour and bread were more natural, and that the country people who ate it lived longer and were healthier than those who did not — referring to country people who had lived in the past. (It was also claimed that whole-wheat bread prevented constipation. I'll deal with this claim later in the chapter when I discuss high-fiber bread.)

By and large, manufacturers try to provide what they think people want, that is, what they think people will buy. Now that the demand for whole-wheat bread has risen, it is being supplied in ever-increasing quantity.

As we learned more about the actual nutrients found in wheat, it certainly did become clear that most of the thiamine, niacin, and iron present in whole-grain flour was removed during the production of white flour. Along with the nutrients, most of the fiber, which has turned out to be very important, was also removed.

What is flour? The wheat kernel consists of a central portion, which is white, called the endosperm; a series of outer layers; bran; and a flake of high-protein, high-fat material, the wheat germ. The bulk of vitamins and minerals are in the outer layers, especially in the wheat germ. White flour is made primarily from the endosperm; whole wheat, from the entire wheat kernel. The white endosperm is relatively bland tasting, while the outer layers have a clearly distinguishable grainy flavor.

After World War II, experiments were carried out on German children in orphanages. These children were fed white and whole-grain flour, with and without enrichment, and were getting most of their protein from wheat, with a little milk added. The results showed no differences in the children's rate of weight gain at all and showed that all achieved "an excellent state of general nutrition."*

*W.R. Aykroyd and Joyce Doughty, *Wheat in Human Nutrition*, FAO Nutritional Studies: No. 23 (New York: Unipub, 1970), p. 107.

While there is disagreement on the reasons for this, the experiments "undeniably show that wheat flour in any form is good nutrition." There is no question that more nutrients are removed than are restored by enrichment, but there is no evidence that this matters nutritionally, when we consider that bread is such a small part of our diets. (Obviously, when we consumed more than 200 pounds of flour per person per year, as we did around 1910, we were more dependent on the vitamins in flour than we are now, when we consume only about 100 pounds per person per year.) In fact, this view is so strongly held in other countries that enrichment is not allowed in France and is considered unnecessary in Holland. In England, the amounts of nutrients added back to flour are less than the amounts used in the United States.

The idea underlying our original U.S. enrichment program was that our diets were varied, we had many sources of protein, vitamins, and minerals, and that we were not trying to make bread a complete food, sufficient to sustain life all by itself.

It was evident, however, back in 1941, when our voluntary enrichment program began, that flour provided an excellent medium to provide thiamine, niacin, riboflavin, and iron with no harmful effect on the population. At that time, our daily average consumption of bread was six ounces per person per day (137 pounds per person per year). Basically, the amount of thiamine, niacin, and iron added was meant to restore those nutrients to the original whole-wheat level. Riboflavin was added at higher levels than in whole wheat because it was judged to be necessary in greater quantity.

Pellagra was a fairly common disease in the South in the 1940s, and the introduction of enrichment to white bread played a major part in virtually wiping it out.

Production of Flour

The modern system of milling wheat was invented in Hungary and has been in use for over 100 years. The slightly moistened wheat berry is gently crushed between rollers to break the kernel and is then sifted to remove any flour released. This is repeated many times in order to minimize the breaking of the bran and germ, thus "contaminating" the white flour with brown material.

In stone grinding, the kernels are crushed between millstones. Proponents of stone-ground flour claim that the vitamin content is higher than steel-roller-milled flour because the kernel isn't heated as much. (High-speed milling produces slight elevation of kernel temperature because of friction created in grinding.) There is no evidence of nutritional differences between stone-ground flour and flour otherwise milled. Stone grinding is simply a somewhat more expensive way to grind wheat and is used to justify an enormous difference in the price of flour for the consumer.

Wheat flour is the only substance in all of nature that can be mixed with other ingredients, such as milk, sugar, fat, and eggs, to form a dough in which the gas that raises it is retained until the protein (gluten) in the flour sets under heat. This keeps the bread light, expanded, soft, and pleasant to eat. Dough made from other flours lose gas and tend to disintegrate, making baked products made from pure rye flour or corn flour very dense, as in Westphalian pumpernickel.

This ability to form a dough that holds together, holds the gas, expands without collapsing, rises in the oven, and, at the peak of its rise, sets and holds its size, is part of what is called the "baking quality" of flour. Other factors that enter into the baking quality are how sticky the dough gets, how much water it takes to make a perfect dough, how fast the yeast rises (which depends on the enzyme content of the flour), how well the dough bakes out, and whether it remains gummy and pasty even after baking.

In the case of industrial and commercial baking, the additional element of timing, as mentioned earlier, is critical. It is this need for controlled baking quality that makes necessary many of the materials added to bread products, particularly for the bleaching and bromating of flour.

After flour is milled, it undergoes change during storage. The oxygen in the air reacts with the yellow flour pigments and tends to whiten the flour; in addition, the flour's ability to form strong, pliable, gas-holding dough is improved through aging.

The general practice in industrial baking for many years was to buy bagged flour (in 100-pound cloth sacks) and store it for as long as a month or more. Many millers also maintained stocks of aged flour that they supplied to bakers. This practice required considerable investment of storage facilities and handling labor, and greatly exposed the flour to contamination by rats, mice, and insects. Fumigation of stored flour was not only common, but virtually necessary to kill insect eggs and larvae.

Another problem with aging flour to improve its baking quality is that the aging requirements are not constant: each wheat variety, each crop, nearly each milling batch requires a different amount of time for aging. Therefore, bakers are continually compensating for variations in the baking quality of flour.

The combination of economic pressures — increasing costs of space and handling labor — the need for consistent quality and more sanitary conditions, as well as the desire to improve consumer acceptance of baked products, led to efforts to bleach flour to accelerate aging.

In bleaching flour to be leavened with yeast, a number of agents are permitted "in a quantity not more than sufficient for bleaching and artificial aging effect" (Code of Federal Regulations). The only labeling requirement is the use of the word "bleached"; the name of the bleaching agent is not required. Those agents include oxides of nitrogen, chlorine, chlorine dioxide, chlorine, benzoyl peroxide.

The purpose of bleaching flour used for cakes that are chemically leavened is primarily to whiten it. Bleaching also weakens the gluten in flour, which allows the carbon dioxide to diffuse throughout the batter and bake off, rather than remaining trapped in the cake. Thus the batter rises evenly, instead of forming a mountainlike cake — low on the sides and high in the middle. Only bleached cake flour permits cakes to be baked that contain more sugar than flour — so-called high-ratio cakes.

Bleached flour is not permitted in other countries, such as Britain and Japan. Extremely sweet cakes such as we know are therefore not produced in those countries.

It is quite likely that as we learn more about sugar in our diets and become more concerned about reducing it, we will see more cakes with unbleached flour and less sugar.

There are already a great many bread products with unbleached flour, where bakers are prepared to take more pains to compensate for variation, and/or settle for a slightly denser loaf of bread, in response to consumer demand for unbleached flour products.

□ CONTROL □
The Reason for Most Additives

In all industrial baking, it is necessary to have a high degree of predictability of all characteristics of the product: size, shape, weight, flavor, color, keeping quality. In order to achieve this, control of ingredient properties and process conditions is critical. Since so many of the ingredients in baked foods are natural, and nature is fickle, these ingredients may vary enormously in properties that are important to the baking process. Even in small-scale retail baking, there must be reasonable control, or else the bread may not be baked when the customers arrive.

A large-scale bakery may use an oven that produces 5000 loaves per hour — more than one every second. This means that any problem that causes a delay causes a backup. Some additives are used to minimize the chances of such delays occurring.

■ ENZYMES

As you will recall from the beginning of this chapter, the starch in the flour must be converted into a form the yeast can act upon, and the enzyme amylase serves that purpose. This enzyme is often missing in flour, but more important, the amount present varies, which means that the fermentation time for yeast would vary. How would you like to operate a bakery and never be able to predict when the dough is ready to go into the oven? It would be impossible to manage. Therefore, it is absolutely necessary to compensate for enzyme deficiency and to standardize what is called the "gassing power" of flour by adding enzymes in the form of malted wheat flour, malted barley flour, or an enzyme extracted from a mold: alpha amylase. (Malting is a "natural" way of producing the needed enzymes, by moistening grain and permitting the grain to germinate.)

■ DOUGH CONDITIONERS

The mechanical production of bread in high-speed bakeries requires that the dough behave in a uniform fashion so that machines can be adjusted to divide the dough into uniform weight pieces that rise in a uniform time and bake to a uniform color in a uniform baking time.

To help control dough properties, calcium peroxide, potassium bromate, and calcium stearoyl-2-lactylate are used. These substances, used in minute quantities, prevent bread dough from sticking to machinery, or make the dough expand more in baking.

■ OXIDIZING AGENTS

Potassium bromate or iodate, or calcium bromate or iodate, may be added to flour to the extent of 50 parts to over 1,000,000 parts of flour. These substances help to make bread larger for the weight of dough used and help to give bread a "better" texture: thinner cell walls, softer. In any bakery product, the quantity permitted is not more than 75 parts in 1,000,000, which permits a bit more oxidizing agent to be added to the dough. These oxidizing agents are classified as food additives.

■ PRESERVATIVES

Many breads now claim "no preservatives." Many people think that preservatives keep bread soft, which is not the case.

Rather, these preservatives delay the molding of bread and the formation of "rope"—a bacterial growth that looks like strings inside the loaf. Such mold growth occurs in wrapped bread in the home, if it occurs at all. It is caused by the high-moisture content of bread, plus ever-present mold spores in the air.

The preservatives used in breads and other bakery products are calcium propionate, sodium propionate, occasionally vinegar (acetic acid), and sodium diacetate—a dry form of vinegar, since it forms acetic acid in the dough. The materials are GRAS. The propionates are found in Swiss cheese and in many foods in nature.

Bread that is baked with a crisp crust and baked for a longer time and not wrapped is less likely to mold, and consequently such preservatives are not used.

Other preservatives are used in other bakery products: sodium benzoate in fruit fillings, sorbic acid or potassium sorbate in cake products and doughnuts, as well as in fillings, toppings, and icings—all for the purpose of preventing or delaying mold growth.

■ EMULSIFIERS

The products that make baked products stay "fresh"—soft and moist—longer are the emulsifiers.

In cakes, emulsifiers are also used to produce a very fine cell structure. Since egg yolk acts as an emulsifier, added emulsifiers reduce the amount of egg necessary. They also assist in making the baked product fluffier—bigger for the same weight.

For most bread sold in the United States, the "squeeze test" is the consumer's method for determining the freshness of the bread. The more air there is in it, the easier it can be squeezed. Emulsifiers, by promoting high air content, contribute to softness.

Factory-produced bread does not get to the supermarket shelf until the day after it is made. It is picked up from the market after another two days and then sent to a thrift store where it may be offered for sale for another three to four days.

This would be impossible without emulsifiers. The bread would be hard in two days. The consequence of this would be the waste of a fair amount of bread or the deliberate production of less so that the bread would sell out every day. In less-populated areas, and in cities distant from the bakery, deliveries would have to be more frequent than presently — once every two to three days. Delivery costs are already a very large part of the cost of baked foods — sometimes larger than the ingredient cost to industrial bakers — and increases in those costs would be passed on to the consumer, resulting in higher prices for bread.

Cakes have a longer shelf life because of the sugar and shortening and eggs: the prime purpose of emulsifiers is to reduce the amount of fat and eggs required, which results in cost savings and products more in accordance with today's dietary guidelines — fewer calories, less fat.

The emulsifiers used in baked foods are many. Each has special characteristics, but the most common are mono- and diglycerides, polysorbate 60, calcium stearoyl-2-lactylate, sodium stearyl fumarate, ethoxylated mono- and diglycerides, lecithin, hydroxylated lecithin, tartaric acid esters of monoglycerides.

The mono- and diglycerides are GRAS and permitted in other countries as well as the United States. Lecithin is a natural food material extracted from soybeans and is GRAS. The others are additives, a number of which are not permitted in England, France, Germany, and Japan. They are by no means absolutely necessary, merely representing improvement in emulsification and/or costs savings over the mono- and diglycerides.

Today, owing to consumer pressure, an ever-growing variety of products is available with only GRAS or food materials present. Just read the label. Usually those products are more expensive than the additive-loaded products, partly because of costs and partly because there is not enough competition yet among the less sophisticated, lower-additive products.

□ THE COMPOSITION OF BAKED FOODS □

Quite understandably, because of their high sugar and fat content, many baked foods do not voluntarily contain nutritional labels. Most packaged breads are so labeled, but not the sweet, high-fat products, where the nutritional label would be most useful to those who are controlling their diets!

Provided below are two charts, one that shows the amount of sugar, shortening, eggs, milk, and fat used in each type of bakery product; the second showing the comparative carbohydrate, fat, and calorie analyses.

In each category, there is a tremendous variation: for example, a cinnamon roll might be made with 10 percent added shortening or 20 percent added shortening, cakes may vary as much as 20 percent in sugar content, cookies vary in shortening and sugar content.

Sugar, shortening, and eggs produce the richness and tenderness of the baked product. In the category of yeast-raised products, we move from the simplest Italian-type bread containing flour, water, yeast, and salt, to common white bread by adding sugar and shortening, and then progress all the way to Danish pastry by adding more and more sugar, shortening, and eggs.

In the chemically leavened category, we move from pancakes to rich cakes in the same manner. We must consider, however, that by the time we add butter and syrup to our pancakes, we match or surpass the sugar and shortening content of a rich cake.

Muffins — not cupcakes — are probably your best calorie bet in the entire line of sweet baked foods.

If we are to adhere to the new dietary guidelines, it would be extremely helpful to us to have nutritional labels on the sweet, high-calorie baked foods, as well as on bread. In my opinion, such labeling could increase the consumption of such products in the long run, since consumers would be able to ration themselves instead of completely avoiding some of the things they like best. It would also stimulate competition among bakers to produce lower-fat, lower-calorie sweet baked products that are satisfying. This has happened in the cheese industry, where manufacturers are now making low-calorie vegetable-fat cheese.

☐ COMPARATIVE RICHNESS OF BAKED FOODS ☐

Yeast-Raised Products

	% Sugar	% Shortening	% Eggs	% Milk
Italian Bread	0	0	0	0
French Bread	2	1	0	0
Whole-wheat Bread	3–4	2	0	1–2
White Bread	3–4	2–4	0	2–3
English Muffins	0	0.5	0	0–2
Doughnuts (Glazed)	25–30	20–25	0–3	0–3
Hamburger Rolls	2–4	3–6	0–3	3–4
Hard Rolls	0–3	0–2	0	0
Coffee Cake with Icing and Filling	25	10–15	0–10	2–5
Danish Pastry	10–15	30–35	10–15	4

Products vary from bakery to bakery.

Chemically Leavened (Cake) Products

	% Sugar	% Shortening	% Eggs	% Milk
Pancakes	0–3	0–3	0–5	3–5
Muffins	25	10	0–5	2–4
Biscuits	5–7	5–10	0	2–4
Bran Muffins	20	6–10	0–5	2–4
Doughnuts (Plain, Cake Type)	15–20	25	0–5	0–3
Cake (Un-iced)	35–40	15–20	5–10	2–5
Cookies	20–25	20–25	3–15	0–2
Angel Food Cake	35	0	25–30 (whites)	0

□ ANALYSIS OF BAKERY PRODUCTS □

Yeast-Raised Products

	CALORIES/OUNCE	% FAT	% CARBOHYDRATE
Italian Bread	80	0–0.5	56
French Bread	83	1–2	55
White Bread	73	3–4	51
Whole-wheat Bread	70	3–4	48
English Muffins	70	1	46
Hamburger Rolls	85	5–6	53
Coffee Cake with Filling and Icing	92	9–10	49
Danish Pastry	122	25	46
Hard Rolls	90	3	60
Doughnuts	120	27	37

Chemically Leavened (Cake) Products

	CALORIES/OUNCE	% FAT	% CARBOHYDRATE
Biscuits	107	17	46
Doughnuts (Plain, Cake Type)	113	19	51
Angel Food Cake	73	0	60
Plain Cake (iced)	106	14	60
Cookies	150	20+	60+
Muffins	85	10	42
Pancakes (made with milk)	100	6	32

□ UNDERSTANDING BREAD LABELS □

Let's compare the labels of two white breads, which we'll call Brand A and Brand B.

Brand A Ingredients

Enriched flour (barley malt, iron [ferrous sulfate], niacin [a B vitamin], thiamine mononitrate [B_1], riboflavin [B_2]), water, sugar, partially hydrogenated vegetable and/or animal shortening (may contain soybean oil and/or lard), yeast, salt, soy flour, calcium sulfate, whey, dough conditioners (may contain sodium stearoyl-2-lactylate, dicalcium phosphate, mono- and diglycerides, monocalcium phosphate, or potassium bromate), calcium propionate (to retard spoilage).

Brand B Ingredients

Unbleached enriched spring-wheat flour (flour, malted barley flour, niacin [a B vitamin], reduced iron, thiamine mononitrate [B_1], riboflavin [B_2]), water, corn syrup, partially hydrogenated vegetable shortening (soybean oil), nonfat milk, salt, fresh yeast, grade-A creamery butter, golden honey, mono- and diglycerides, calcium propionate added to retard spoilage.

One says "enriched flour," which we must assume is bleached, since no claim is made that it is unbleached, although the law requires that the label state whether the flour is bleached or unbleached. The other proudly claims unbleached spring-wheat flour. Milk versus whey; vegetable shortening versus hydrogenated lard; butter and honey in one, not in the other; sodium stearoyl-2-lactylate in only one. Both breads contain calcium propionate as a preservative.

From the ingredient listing, Brand B seems to be more wholesome — and costs 20 percent more.

You ought to remember that bread usually contains less than 1 percent salt. Since by law ingredients must be listed in order of the quantity present, you know that the amounts of butter and honey in Brand B must be less than the amount of salt, that is, less than 1 percent — and probably considerably less.

Now compare the nutritional labels.

Brand A

Nutritional Information per Serving

Serving size 2 ounces (approx. 2 slices)
Servings per package 8

	Per 2-ounce serving (rounded figures)	*Per 6-ounce daily*
Calories	150	440
Protein	5 grams	14 grams
Carbohydrates	27 grams	82 grams
Fat	2 grams	6 grams

Percentages of U.S. Recommended Daily Allowances (RDA)

	Per 2-ounce serving (rounded figures)	*Per 6-ounce daily*
Protein	8	20
Vitamin A	0	0
Vitamin C	0	0
Thiamine	15	45
Riboflavin	8	25
Niacin	10	30
Calcium	6	20
Iron	8	25

Brand B

Nutritional Information per Serving

Serving size	2 slices	Protein	3 grams
Servings per container	10	Carbohydrates	20 grams
Calories per serving	130	Fat	3 grams

Percentages of U.S. Recommended Daily Allowances (RDA)

Protein	4	Riboflavin	6
Vitamin A	*	Niacin	6
Vitamin C	*	Calcium	4
Thiamine	8	Iron	8

*Contains less than 2 percent of the U.S. RDA of these nutrients.

The first thing to look at is the serving size. Brand A says 2 ounces, *approximately* two slices. Brand B doesn't tell you how much the two slices weigh, but there are twenty in the package, and the package weighs 16 ounces, so two slices weigh only 1.6 ounces. While it is true that Brand B has fewer calories in a portion (two slices), it is actually higher in calories per ounce than Brand A. Still, we eat bread by the slice, not by the pound, so if we are watching our calories, we would reduce our intake with Brand B. However, Brand A supplies more protein, less fat, and a greater contribution of vitamins and minerals than Brand B.

Whole-wheat breads may vary in additive content just as much as white breads, so we won't repeat the comparison of ingredients. The most interesting comparison here is among the nutritional analyses. The chart below compares the calories, vitamins, and minerals (percentage RDA) in servings of a white, a whole-wheat, and a wheat-germ bread.

	White Bread	Whole Wheat	Wheat Germ
Calories	150	140	120
Vitamin A	0	0	0
Vitamin C	0	0	0
Thiamine	15	8	10
Riboflavin	8	2	10
Niacin	10	5	10
Calcium	6	2	4
Iron	6	8	8

So, a general statement can be made to summarize the situation: Whole-wheat bread does not supply as much of the known vitamins as white enriched bread; wheat-germ bread makes up for the enrichment. If there is any advantage to eating whole-wheat or wheat-germ bread rather than enriched white bread, it cannot be because of the known minerals or vitamins, but must be for some other reason. This brings us to the matter of fiber.

■ HIGH-FIBER BREAD

Perhaps the most important development in the whole-wheat-bread–white-bread controversy is the discovery that there seems to be a very direct link between the amount of cereal fiber in our diet and the incidence of bowel diseases, including bowel cancer and diverticulosis.

The evidence seems to be overwhelming. Not only are these illnesses more widespread where cereal fiber in the diet is low, while virtually nonexistent where the diet has plentiful cereal fiber, but the evidence shows that the addition of bran to the diet positively helps to relieve diverticulosis. This evidence is so strong that the treatment for diverticulosis has done a complete about-face: for fifty years, the treatment was to eat a low-fiber diet; today, it is to add bran to the diet, with better results than the surgery that doctors were driven to perform to alleviate this painful disease. Furthermore, evidence shows that the addition of bran to the diet keeps the bowel healthy.

If all we did was replace our white-flour foods with whole-wheat foods, we would have sufficient cereal fiber. Or we can simply add bran in other ways, via bran cereals or bran sprinkled on cereals or through one of the high-fiber breads.

There are basically two kinds of such breads: white breads with added wood fiber, and dark breads with added cereal fiber. The dark breads are often called "natural-fiber breads." For those who want white bread and toast with high fiber, the situation is more difficult. While such breads are available with wood fiber, recent evidence shows that this source is not desirable, and in fact brands that were very popular have virtually disappeared.

A confusing element in the high-fiber white breads using wood cellulose is the low-calorie claim. The reason for this seems to be the higher water content in these breads. In fact, they contain more water than flour, as is evident in the ingredient listing below.

Compare the ingredient listings on three brands of high-fiber bread. Brand X is a white bread and sells for 75¢ a pound. Brands Y and Z are dark breads called "natural-fiber bread."

Brand Y sells for 79¢ a pound; Brand Z is 79¢ for 18 ounces, or 70¢ a pound.

Brand X

7.5 percent fiber
"400 percent more fiber than whole-wheat bread"
INGREDIENTS: Water, flour, powdered cellulose, wheat gluten, brown sugar, salt, sugar, yeast, lactalbumin, calcium sulfate, dough conditioners (may contain sodium stearoyl lactate, mono- and diglycerides, polyglycerate 60 [ethoxylated mono- and diglycerides], polysorbate 60, potassium bromate, barley malt), artificial flavor, calcium propionate as a preservative . . . [and the usual enrichment materials].

Brand Y

1.7 percent fiber
Natural Fiber Bread
INGREDIENTS: Unbleached enriched wheat flour, water, coarse-ground whole-grain cereals (wheat, oats, corn, rye), unprocessed millers' bran, golden honey, yeast, salt, raisin syrup, coconut, yeast nutrients (calcium sulfate, ammonium sulfate, potassium bromate), sesame seeds, mono- and diglycerides. No preservatives added.

Brand Z

2.1 percent fiber
Natural Fiber Bread
INGREDIENTS: Flour, unbleached, stone-ground whole-wheat flour, water, brown sugar, wheat-berry bran, ground soybean hulls, honey solids, 2 percent or less of each of the following: partially hydrogenated soy oil, yeast, salt, vinegar, malted barley flour, ascorbic acid for uniform bread texture . . . [plus the usual enrichment materials].

There are two interesting things to be learned from this comparison:

1. There is no shortage of breads with fiber. Once it was clear that there would be a consumer demand, fiber breads were created to satisfy every kind of consumer demand, from white, airy, and light to dark, heavy breads.

2. If you want white bread *without* additives but with fiber, you are out of luck. The additives are necessary to produce the light, airy loaf.

Some new materials turn up in the labels we've just looked at. Wheat gluten is the separated protein of wheat that is added in small quantities (under 5 percent) to some baked products in order to strengthen the flour. In Brand X, where so much water is used (note: it is the first ingredient listed), wheat gluten helps to give the bread some structural strength and therefore makes it able to hold the air and end up as a big loaf for the weight. Lactalbumin is a protein separated from milk. Barley malt is shown as a dough conditioner. It contains enzymes (*all* enzymes are natural — man hasn't been able to synthesize any yet), which are needed for the bread dough to ferment properly.

■ SPECIALTY OR VARIETY BREADS

Two trends gave birth to a tremendous proliferation of bread varieties: consumer discontent with the flavor of and additives in factory-made white bread; and the introduction of the continuous-mix process to the production of white bread. The continuous-mix process was originally so mechanically rigid that it was virtually impossible to produce any varieties, but those bakers who introduced the process have since modified it or abandoned it because consumers were dissatisfied with the loss of flavor through the process and wanted different types of bread. Increasing demand for whole-wheat bread, more natural breads, and dark breads has made variety breads one of the few growing areas of the entire baking business.

There are some things interesting to know about such breads. For example, what is rye bread? The one thing it is *not* is bread made from only rye flour or even mainly rye flour, nor is corn bread made with corn flour primarily, nor oat bread with oat flour. Common rye bread is made with wheat flour and rye flour and is flavored with a rye sour (a ferment of yeast, rye flour, and sometimes milk, resulting in the production of lactic acid). Real sour rye is actually fermented with a rye sour. Commercially, it is more common for the rye sour to be produced by a flavor manufacturer and sold to the baker for use in his rye-bread formula.

The amount of rye flour is about 15 percent of the total flour. As the percentage of rye flour is increased, the bread becomes denser and denser, culminating in whole-rye pumpernickel, which is a bread that is almost totally unexpanded. What is usually sold as pumpernickel is actually the same as common rye bread, but is colored with caramel color.

Rye bread is not usually enriched and consequently is lower in vitamins and minerals than white bread. The nutritional label shows no important difference in analysis otherwise.

There are a number of specialty breads offered for the purpose of losing weight. Usually, the lower-calorie breads are such mainly because of the slice size and weight. You can buy bread with as little as 35 calories per slice as compared with 70 calories for a standard white slice. The enrichment provided is proportional: the lower the weight of the slice, the lower the vitamin and mineral level. This, however, should in no way detract from the value of being able to make a two-sided sandwich from two slices of bread for the same number of calories that ordinarily would be provided in only one slice. The only thing to think about is whether you want to pay a premium price to the baker to help you eat half as much — which may very well be worth it!

One nationally advertised specialty bread contains bran, whole rye, soybeans, flaxseed meal, in addition to bleached wheat flour and other common bread ingredients — with a full complement of additives. The nutritional analysis shows no significant difference from white bread, although vitamin and iron content is slightly higher (but the slices are slightly heavier too).

Another brand is sold as a special-formula dietary bread with slightly fewer calories per slice — 55 versus the usual 65 to 70 — because of slice weight and lower fat. This brand also sells at a premium over ordinary white bread, but still contains all the additives.

Question: Which type of bread should you give to your children, and why? Here are a few guidelines.

Using This Information to Select Bread

1. If you object to additives, there are plenty of brands without additives, using only GRAS materials.
2. There is probably not a better nutritional buy than factory-produced enriched white bread — if you don't mind the taste. For about 6 cents or less per serving, you can obtain significant amounts of vitamins, minerals, a fair amount of protein, with the calories in good balance, with less than 20 percent of the calories coming from fat (if you don't butter the bread).
3. Before you buy a bread priced higher than ordinary white bread, read the label to make sure you are getting what you think — unless you buy it for taste — and compare it with the white-bread label.
4. Man does not live by bread alone. You could do a lot worse than picking the bread you buy for the taste you like the best, if you can afford the price.
5. I would prefer for my children to eat any enriched bread or whole-wheat bread rather than any specialty bread, since specialty breads tend to be lower in nutrients.
6. Check ingredients on the label.

□ SUBSTANCES ADDED TO BREAD □

MATERIAL ADDED	LEGAL STATUS (and GRAS class)	PURPOSE
Ammonium sulfate	GRAS (1)	Production
Ascorbic acid	GRAS (1)	Production
Barley malt	GRAS (food)	Production
Calcium propionate	GRAS (1)	Preservative
Calcium sulfate	GRAS (1)	Production
Caramel color	GRAS (1)	Coloring
Cellulose	GRAS (1)	Nutrition
Dicalcium phosphate	GRAS (1)	Nutrition/ dough conditioner

Material Added	Legal Status (and GRAS class)	Purpose
Ethoxylated mono- and diglyceride	Additive	Increased shelf life
Ferrous sulfate	GRAS (2)	Nutrition
Gluten flour	Food	Production
Lactalbumin	GRAS (food)	Nutrition
Lecithin	GRAS (1)	Increased shelf life
Mono- and diglycerides	GRAS (1)	Increased shelf life
Monocalcium phosphate	GRAS (1)	Yeast food/ dough conditioner
Niacin	GRAS (1)	Nutrition
Polysorbate 60	Additive	Increased shelf life
Polysorbate 80	Additive	Increased shelf life
Potassium bromate	Additive	Production
Potassium iodate	GRAS (1)	Production
Reduced iron	GRAS (2)	Nutrition
Riboflavin	GRAS (1)	Nutrition
Rye-sour flavor	GRAS (not classified)	Flavoring
Sodium diacetate	GRAS (1)	Preservative
Sodium stearoyl-2-lactylate	Additive	Production
Thiamine/ mononitrate	GRAS (1)	Nutrition
Vinegar	GRAS (1)	Preservative
Wheat gluten	Food	Nutrition and production
Whey	GRAS (milk component)	Partial milk substitute

□ SWEET BAKED FOODS □

While nonsweet baked foods provide nutritional labeling, there is a notable absence of such information on the sweet products. Such labeling, at least for the time being, is optional. Efforts are being made to institute mandatory nutritional labeling, but it may take a few years to achieve.

A good rule of thumb for baked products is this: *If there is no nutritional label, it is probably high in sugar, fat, and calories.*

This is not always the case; there are other reasons why a product may not be nutritionally labeled. For example, if angel food cake (which is high in sugar, but contains no fat) were labeled, we would all wonder why the *other* cake products were *not* labeled! The baker is hesitant to open this Pandora's box. Yet, for those who should restrict their intake of fat, sugar, and calories, the products that need the labels most don't have them.

A recent change in cake composition has been the introduction of enriched flour. As per capita consumption of bread declined, the FDA and other health authorities were concerned that the diet would not supply the necessary amount of vitamins and minerals that had been provided in enriched bread. Therefore, on a voluntary basis, the baking industry began to use enriched flour in cake on the advice of the government health and nutrition experts.

As in the case of bread, the fact that a product is made in a retail bakeshop does not mean that the ingredients are any different from those of cakes made in a large factory. You just can't tell because there is no label. Retail bakers use many of the same materials used by industrial bakers.

Cake-product labels are much more complex in their ingredient statements than bread products. This is true of most sweet bakery products, since they usually contain icing or coating of some kind, and fillings as well. These coatings and fillings tend to dry out, sweat, run, mold, stain, suck moisture from the crumbs or the baked product, dull, turn unsightly, or dissolve, and need special materials to help stability.

Often the filling, such as a fruit filling, is produced outside the bakery and must be preserved during storage and transit. An example is the jelly in jelly doughnuts, which often contains sodium benzoate, not for the sake of preserving the doughnut, but so that the filling stays uncontaminated by yeast or mold until the baker is ready to use it.

The very act of packaging creates a need for ingredients that might otherwise be unnecessary; for example, substances added to prevent icing from sticking to a package.

The shipment of baked goods from central bakeries to distant points — often hundreds of miles away — requires additional ingredients; for example, starch or other thickening agents in pie filling so that the pie doesn't leak in transit.

The shelf life of sweet baked foods, especially chemically leavened ones, is necessarily quite long — sometimes more than ten days — and the product must be protected against all the changes that might occur in that time period to keep it edible.

As we contemplate all these considerations, we may wonder whether there is any nutritional sense to the manufacture of commercial sweet baked goods. Of course, we do not rely on such foods for nourishment, rather we must be able to *enjoy* some foods that may not contribute much besides some calories and, now, with enriched flour, a small share of some vitamins and minerals.

If bakers are smart, they will be searching for new methods and recipes for sweet baked foods in which the calories are not quite as empty as they are now, and with significantly less fat.

There is only one piece of practical advice with respect to this class of foods: *use moderation,* and assume that anything you buy without a nutritional label will contain 250 to 300 calories in the portion you eat — which will be about 2 ounces — and will contribute 12 to 15 grams of fat, a full 16 percent of the amount of fat that you should be eating in a day.

Be sure to figure the butter or jam you add to anything, like a muffin or pancake, when you think about the fat and calories in it.

Finally, there is not too much point in trying to discriminate between sweet baked foods nutritionally. Eat what you like best — in moderation.

■ THE LABELS ON SWEET BAKED FOODS

Very commonly you will find the phrase "may contain beef fat and/or lard and soybean and/or cottonseed and/or palm oil" in parentheses following the words "partially hydrogenated animal and vegetable shortening." Quite naturally one wonders, What do they mean by "*may* contain"? Don't they know what they put in? The reason for this phrase is economic. The prices of the different fats vary constantly, and shortening manufacturers blend the oils to attempt to maintain a reasonable, steady price, or to minimize the cost to the baker, which they must do to stay competitive.

Technology in the shortening business has developed to the point where many different fats can be treated by hydrogenation to produce a final shortening that is suitable for baking. If labels were to be changed every time the source of fat in the shortening were changed, it would drive bakers and consumers out of their minds and require enormous inventories of labels, one for each type of shortening. Hence, the FDA permits the type of labeling shown.

Incidentally, it should be assumed, unless otherwise stated on the label, that partially hydrogenated shortening is saturated fat.

Another natural question is: Why not use unhydrogenated oils — unsaturated fats? The answer is that the products would be greasy. (Some products are made this way and are greasy.) The packaging material would be stained; so would sugar coatings of fried products.

Remember: sweet baked products, at least at present, are not for your health — they are to enjoy.

Now we can summarize the new ingredients you will find on the labels, and their uses.

Whey and sodium caseinate are often used. As the price of dry skim milk climbed higher and higher, manufacturers found it more economical to use its components in different proportions than normally found in the skim milk.

Sorbic acid is commonly used in cakes, sometimes in the form of potassium sorbate. This mold inhibitor (preservative) cannot be used in yeast-leavened doughs because it interferes with the growth of the yeast, which produces the carbon dioxide to raise the dough. This is not a problem in chemically leavened cakes. Sorbic acid prevents mold growth especially on the surface in high-moisture-content packaged cakes.

Artificial flavors are used for two reasons in many cakes:

1. They are much cheaper than natural flavors.

2. They can be made much stronger than natural flavors and, as in the case of vanilla, do not bake out as readily.

The labels are quite clear when it comes to this ingredient, and it is possible to buy baked foods of every type with natural flavors, which usually are, and should be, more expensive. The artificial flavors are often blends of compounds that have been found in natural foods, except that the compounds are made synthetically and are combined in different proportions than in nature, with some of the flavor components missing, which explains the taste difference.

Fortified eggs is another new ingredient for us. It simply consists of whole eggs with extra egg yolk added for baking quality.

Modified food starch is starch that has been bleached with chlorine or otherwise treated chemically in one of more than twenty ways — much too complex to describe here. All modified food starch is an additive, not GRAS.

Sodium propionate is another form of the preservative against mold, calcium propionate.

Fumaric acid is an additive, not GRAS, although it is found in many plants and is necessary to vegetable and animal tissue respiration. It provides a tart flavor of a specific type and serves as an antioxidant. It is produced commercially by the action of a fungus on glucose.

BHT and BHA, added to preserve freshness, are antioxidants used in frying fats and other fats to prevent them from becoming rancid. They are GRAS, class 3, when used in fats and oils at a level of less than 0.02 percent. Otherwise they are additives. The hypothesis has been stated that the decline in stomach cancer in the United States might conceivably be due to the introduction of antioxidants into fats — with no proof yet.

Rancid fat is not healthful. While antioxidants are not necessary to assure fresh fat (which can be obtained with greater saturation of the fat or the use of hydrogenated fats, which are more stable than those usually used for frying), BHA and BHT make feasible the use of lower-priced fats in certain operations.

If a baker were willing to throw out used fat every day, or if he fried enough daily to use up most of the fat each time, BHA and BHT would be unnecessary; but these measures are costly and impractical. The use of BHA and BHT protects us against rancid fat, but again, there are products available without it, as you can tell from the label.

Glycerin is used to retain moistness and is digested by the body similarly to alcohol. Indeed, glycerin is an alcohol. It is GRAS.

Sodium aluminum phosphate is a baking acid preferred for some uses because of its reaction rate and/or taste.

Gelatinized wheat starch is wheat starch that has been precooked and dried, that is all. It is a thickener and modifies baking quality of doughnuts and cakes.

Rye flour in cakes or doughnuts is present as a cost reducer. At times rye flour is cheaper than wheat flour and a certain amount can be blended with the wheat flour without affecting the quality of the product very much.

Calcium lactate and calcium carbonate are both GRAS; indeed, they are normal body substances as digested. Their purpose is obscure, probably to modify batter consistency.

Calcium sulfate appears on some labels, as does *calcium oxide.* Both are GRAS. Calcium oxide is used to regulate acidity; calcium sulfate to modify batter consistency.

■ FROZEN SWEET BAKED FOODS

Many of the same ingredients as in fresh products are found in such foods, except that there is no need for the use of any mold inhibitors unless prolonged shelf life after thawing is expected.

However, because frozen foods thaw and refreeze so many times between the time of manufacture and time of use, and because home-freezer, as well as store-freezer, conditions are so variable, there are generally more texture modifiers (gums and starches) used to maintain the textural integrity of the product.

Frozen baked products can dry out very readily in the

freezer at home if the package is not sealed with moisture-proof material such as polyethylene. Frozen baked foods will turn stale if held at 20° F or higher. The ideal storage temperature is 0 to 10° F.

Frozen baked foods should require fewer additives to maintain freshness, since that is the very purpose of freezing. Compare, for instance, the following ingredients from frozen and fresh sweet rolls.

Frozen Brand

INGREDIENTS: Enriched flour (with malted barley flour, niacin, iron, thiamine mononitrate [vitamin B_1], riboflavin), skim milk, sugar, fresh whole eggs, butter, partially hydrogenated vegetable shortening (soybean and cottonseed oil), raisins, corn syrup, yeast, mono- and diglycerides, dextrin, salt, modified food starch, cinnamon, dates, dried apples, xanthan gum, vanilla, honey, annatto extract.

Fresh Brand

INGREDIENTS: Enriched bleached flour (flour, niacin, ferrous sulfate, thiamine mononitrate, riboflavin), sugar, vegetable shortening (may contain soy oil and/or cottonseed oil and/or palm oil, which may be partially hydrogenated), water, cherries, corn syrup, whey, yeast, whole eggs, egg white, modified cornstarch, nonfat milk, salt, cellulose gum, mono- and diglycerides, citric acid, dextrose, soy protein isolate, sodium stearoyl-2-lactylate, natural and artificial color, lecithin, ammonium sulfate, sodium caseinate, calcium caseinate, calcium oxide, sodium phosphate, potassium bromate, barley malt.

Mono- and diglycerides produce an airy and fluffy product and permit reduction of fat content, but there is no excuse for the use of polysorbates.

Artificial color includes: yellow for crumb color, caramel for dark brown, titanium dioxide for white coatings, and so forth. While caramel color will usually be named, the others are not.

■ COOKIES

Nowhere is there a clearer choice of products than in the selection of cookies. There are varieties available ranging from the almost completely synthetic to those free of additives and artificial flavors and colors; from those using animal fats to those using vegetable shortening. Here are ingredients from three different labels:

Brand A

INGREDIENTS: Bleached wheat flour, sugar, water, vegetable shortening (partially hydrogenated soya and fully hydrogenated cottonseed and/or palm oil), oatmeal, raisins, corn syrup, nonfat milk, eggs, malt, spice; contains 2 percent or less of each of the following: leavener (sodium bicarbonate, sodium chloride, tricalcium phosphate), salt, artificial and natural flavoring; filled with: dates, water, sugar, cornstarch, salt.

Brand B

INGREDIENTS: Unbleached wheat flour, partially hydrogenated vegetable shortening (soybean and/or cottonseed and coconut oils), raisins, oatmeal, sugar, whole eggs, unsulphured molasses, sugar syrup, baking soda, salt, and vanilla extract.

Brand C

INGREDIENTS: Sugar, vegetable and animal shortening (hydrogenated palm-kernel oil, partially hydrogenated soybean oil and/or lard and/or palm oil), enriched wheat flour (contains niacin, reduced iron, thiamine mononitrate [vitamin B_1], riboflavin [vitamin B_2]), cocoa (processed with alkali), whey, dextrose, corn sweetener, corn flour, cornstarch, chocolate, sodium bicarbonate, sorbitan monostearate, salt, lecithin and polysorbate 60 (emulsifiers), peppermint oil, and artificial flavor.

The choices that are *not* available are low-sugar, low-fat cookies with nutritional labels. Some cookies, however, are lower in fat and calories than others, which you can just about tell by how they "bite." The tougher they bite, the less fat they contain. This is true of ginger snaps and molasses cookies, which contain only 5 to 10 percent fat as compared to 25 to 30 percent fat in the shorter cookies. Oatmeal with raisins is intermediate at 15 percent fat. Raisin cookies and fig bars are also relatively low in fat.

The best-tasting cookies are those made without artificial flavors or colors. They are also the most expensive. It would seem that the wisest thing to do about cookies is to eat better cookies less often and thus kill two birds with one stone — cut down on empty calories and enjoy a better product.

BAKING MIXES

The basic ingredients used in dry mixes are the same as in bakery products. There are some special problems that mix manufacturers must deal with: prolonged shelf life of the mix and the tremendous variation in preparation equipment within the homes.

Protection against fat rancidity during the product's shelf life is insured by the use of antioxidants BHA and BHT and sodium bisulfite. For household convenience, the mix must remain free-flowing and lump-free. Materials are added to prevent lumping: sodium silicoaluminate and silicon dioxide.

Flavors used in mixes are almost always artificial. Artificial flavors can be much more concentrated in strength than natural flavors, so much less can be used. Since these flavors are liquids, lumping of mixes that could be caused by excessive liquid is avoided. Another reason is the cost. Artificial flavors are much cheaper than natural flavors — flavor costs per pound of finished product can vary from a fraction of a penny to more than 5 cents. Artificial flavors can be controlled for uniformity more readily, and often are more stable during storage. But natural flavors can be used, with higher cost and often superior quality, with special processing techniques. These costs are passed on to the consumer.

Artificial color is common. However, there are products and brands that avoid it.

Some ingredients appear because they are ingredients of ingredients. For example, a solvent used to produce artificial flavors is propylene gylcol. This may appear on the label because it is an ingredient of the flavor purchased by the mix maker.

A happy fact about many mixes is that they do have a nutritional label. You may find that the calories per serving are considerably less than I have shown for baked products. That is because the mix label is more optimistic about how little you will eat of a good thing than I am, but, nevertheless, the information is there for you to use.

Foods from the grain group are recommended for daily use by *all* nutritionists, and with our growing knowledge of the need to reduce fats in our diets, grain foods are becoming more important.

This is particularly true of the breads and the lower-sugar-and-fat sweet baked foods.

□ SUBSTANCES ADDED TO SWEET BAKED FOODS □

MATERIAL ADDED	LEGAL STATUS *(and GRAS class)*	PURPOSE
Acetic acid	GRAS (1)	Inhibits mold growth
Ammonium sulfate	GRAS (1)	Yeast food
BHA	GRAS (3)*	Antioxidant
BHT	GRAS (3)*	Antioxidant
Bicarbonate of soda	GRAS (1)	Baking powder ingredient
Calcium acid phosphate	GRAS (1)	Baking acid
Calcium bromate	Additive	Oxidizing agent
Calcium carbonate	GRAS (1)	Nutrition
Calcium iodate	GRAS (1)	Oxidizing agent
Calcium lactate	GRAS (1)	Nutrition
Calcium oxide	GRAS (1)	Dough conditioner/ yeast food
Calcium peroxide	GRAS (not classified)	Dough conditioner
Calcium phosphate	GRAS (1)	Yeast food
Calcium propionate	GRAS (1)	Inhibits mold growth
Calcium stearoyl-2-lactylate	Additive	Dough conditioner

Material Added	Legal Status *(and GRAS class)*	Purpose
Calcium sulfate	GRAS (1)	Yeast food
Caramel color	GRAS (1)	Coloring
Cellulose	GRAS (1)	Fiber source
Citric acid	GRAS (1)	Acidifier
Diglycerides	GRAS (1)	Emulsifier (tenderizes and keeps moist)
Ethoxylated mono- and diglycerides	Additive	Emulsifier (softens and keeps fresh)
Fumaric acid	Additive	Acidifier
Glucono delta lactone	GRAS (1)	Baking acid
Glycerin	GRAS (1)	Keeps moist
Hydroxylated lecithin	Additive	Emulsifier
Lactic acid	GRAS (1)	Flavor
Lecithin	GRAS (1)	Emulsifier
Modified food starch*	Additive (1–5)†	Thickener
Monoglycerides	GRAS (1)	Emulsifiers (tenderizes and keeps moist)
Nutmeg, mace and their essential oils	GRAS (3)	Flavoring
Polyglycerate 60	Additive	Emulsifier (keeps bread soft)
Polysorbate 60	Additive	Emulsifier (tenderizes and keeps moist)
Potassium bromate	Additive	Oxidizing agent
Potassium iodate	GRAS (1)	Oxidizing agent
Potassium sorbate	GRAS (1)	Preservative
Propylene glycol	GRAS (1)	Flavor solvent
Salt	GRAS (4)	Flavor
Silicon dioxide	Additive	Keeps mix free-flowing
Sodium acid pyrophosphate	GRAS (1)	Baking acid
Sodium aluminum phosphate	GRAS (1)	Baking acid
Sodium aluminum sulfate	GRAS (1)	Baking acid

Material Added	Legal Status *(and GRAS class)*	Purpose
Sodium bisulfite	GRAS (2)	Preservative
Sodium caseinate	GRAS (1)	Milk substitute
Sodium diacetate	GRAS (1)	Inhibits mold growth
Sodium propionate	GRAS (1)	Preservative
Sodium silico-aluminate	GRAS (1)	Keeps mix free-flowing
Sodium stearyl fumarate	Additive	Dough conditioner
Sorbic acid	GRAS (1)	Preservative
Tartaric acid	GRAS (1)	Baking acid
Tartaric esters of monoglycerides	GRAS (1)	Emulsifiers
Titanium dioxide	Additive	Coloring

*GRAS (3) in fats and oils. Otherwise an additive.
†Modified starches are not now specifically identified on the label. They may be anything from class 1 to 5, depending on which starch is used.

Chapter 8

CEREALS: HOT AND COLD

Every year, we spend more than $10.00 per person — man woman, and child — on hot and cold cereals. Most of this money (80 percent of it) goes for the cold cereals.

The cereal section of the supermarket is one of the most colorful, with dozens of different products for us to choose from. Our choice is usually made by habit or pressure from our children or, more recently, as a result of greater nutritional consciousness. It is impossible for the average consumer to carefully study all the labels of the myriad products carefully before making a choice. In this chapter, we will provide general guidelines for you to use in picking your way through this jungle.

□ A BIRD'S-EYE VIEW OF CEREALS □

First, a perspective on cereals for breakfast. The standard portion size used on all cereal labels, hot or cold, is 1 ounce. In the case of the cold cereals, the standard portion of 1 ounce is added to ½ cup of milk, and the nutritional information is provided for the product with and without the milk.

For the hot cereals, no such addition is shown, and the nutritional information is provided only for 1 ounce of the cereal itself.

If you serve cereal for breakfast, no matter which one it is, hot or cold, oats, wheat, rice, corn, natural or processed, the total calories in that ounce will run from a low of 95 to a high of 110—5 percent of the day's average calorie requirement. With the ½ cup of whole milk, the calories will run to almost 200, about 7 or 8 percent of the day's average calorie intake.

So, from the point of view of calories, it just is not going to make much difference what kind of cereal you eat. The type of milk you use makes much more difference.

□ WHAT ABOUT PROTEIN? □

Protein content is pretty much the same story as calorie content, but with one or two important exceptions. An ounce of most of the cereals—cold or hot—provides 2 to 3 grams of protein (3 percent of the day's requirement). The ½ cup of milk adds 6 grams of protein, which boosts the total to 14 percent of the RDA.

The exceptions are oatmeal, which provides 4 grams of protein in 1 ounce, and Special K, which provides 6 grams in 1 ounce.

All of the labels clearly show the protein contributed and the percentage of the RDA. Cereal, with ½ cup of milk, does provide a good contribution to the day's protein needs. The negative exceptions are the very-high-sugar cereals, with as little as 1 gram of protein per ounce.

□ THE VITAMIN CONTROVERSY □

Surprisingly, the processed, dry cereals are much higher in vitamin content than the so-called natural cereals, as well as the ordinary hot cereals such as oatmeal or wheat cereal. They are all fortified with vitamins and usually supply in a 1-ounce portion 25 percent of the day's requirements of the vitamins used. Some brands go as high as 100 percent of the day's requirements.

The controversy rages around two issues: critics of the dry cereals claim that the vitamins have to be added to cereals because the vitamins originally in the grains were removed through excessive processing. This is not true. A look at the vitamin content of the whole-grain cereals shows that even those are very low in vitamins compared to the fortified cereals; they never did have very high vitamin content and were important sources of some vitamins for us many years ago because we consumed such large quantities of them.

The second argument is that it costs the manufacturers very little to add the vitamins, but the price of the fortified cereal is much higher than the natural cereal. Well, that's true. It certainly is possible to buy vitamins for less money than they cost in fortified cereal. While I will not defend the social value of dry, precooked cereals, they certainly are convenient and attractive to eat, especially to children. We are quite prepared to pay a premium on other things for the work that has already been done for us; for example, we don't criticize clothing manufacturers because they charge us for more than the cost of the yarn they use. And just as we can choose to weave our own cloth, so can we choose to cook our own cereals instead of buying them ready to eat.

The controversy, however, has created a market for some new products, such as a 100-percent-natural cereal, granola. This market has exploded into many hundreds of millions of dollars annually and is a very good example of how faddism can result in a windfall for food manufacturers with no benefit to consumers.

A typical granola cereal provides in 1 ounce just 3 grams of protein compared to 2 in corn flakes; but in addition, 5 grams of fat, which most of us want and need *less* of, compared to no fat in corn flakes. Most cereals provide no fat or only 1 gram of fat. The vitamin levels are about the same as for the ordinary hot cereals, as are the mineral levels. The price is, of course, much higher. So, as a result of the "natural" fad, consumers get a chance to pay a lot more for a lot less!

The situation is just about the same with another 100-percent-natural cereal: 130 calories, 5 grams of fat (and saturated fat at that — coconut oil, no matter how naturally tropical it sounds, is saturated fat), and in addition, sugar is added, albeit brown. There are no mineral claims made for the 100-percent-natural cereal.

In time, people will catch on to the fact that the word *natural* does not necessarily mean that a food is better for us. We must pay attention to its ingredients and nutritional analysis. In this case, the word "natural" means higher calories, higher fat, no more protein, and a lower vitamin content for more money — hardly a bargain.

□ THE HIGH-SUGAR CEREALS □

Because some cereal manufacturers have produced dry cereals with high amounts of sugar, many consumers now believe that all the dry cereals are high in sugar. They are not, as we can easily tell by reading the label.

Here is an example: On the side panel of a box of a well-known sugar-coated cereal, it says:

Carbohydrate Information per 1 Ounce (28.4 g)	
Starch and related carbohydrates	13 g
Sucrose and other sugars	13 g

This means that a serving of this product contains a bit more than 3 level teaspoons of sugar: quite a lot, but at least you know. The cereal is 45 percent sugar (divide 13 by 28.4, the weight of sugar by the weight of an ounce in grams). In this particular case, the other sugars are dextrose from the corn syrup, and sucrose in the molasses declared on the label.

An all-bran product contains 20 percent sugar, calculated in the same way: the label shows 6 grams of sugar in 1 ounce (28.4 grams).

As a rule of thumb, if you do not want to go through the calculations, the sweetened cereals designed to appeal to kids contain 45 to 55 percent sugar (3 to 4 teaspoons per ounce); the old standbys, about 10 percent sugar (less than 1 teaspoon per ounce). This is approximate, but close enough for determining your sugar intake.

□ CEREAL ANALYSIS □

Per Ounce

CEREAL	CALORIES	PROTEIN (*grams*)	CARBOHYDRATE (*grams*)	FAT (*grams*)
Cocoa Krispies	110	1	26	0
Sugar Smacks	110	2	25	0
Wheaties	110	3	23	1
Cap'n Crunch	110	1	24	2
Quaker 100% Natural	130	3	18	5
Wheatena	110	3	21	1
Special K	110	6	21	0
Corn Chex	110	2	25	0
Total	110	2	24	1
Corn Flakes	110	2	25	0
Product 19	110	3	24	0
Kellogg All-Bran	70	3	22	1
Sugar Frosted Flakes	110	2	25	0
Granola	130	3	19	5
Corn Flour	104	2	22	1
Rolled Oats	110	4	19	2
Rice	110	2	22	0
Whole Wheat	95	3	20	1

Sugar information is on most labels as grams per ounce. It is provided voluntarily but may soon be mandatory.

□ WHEAT GERM □

Wheat germ provides the wheat seed with the starting nutrients it needs to germinate, but it is also a rich source of B vitamins and is fairly high in low-quality protein and some essential minerals. It is 20 percent fat.

One ounce of wheat germ, about ¼ cup, supplies 103 calories, 15 percent of the day's protein allowance, 30 percent of the thiamine, and 30 percent of the phosphorus and zinc RDAs — plus lesser amounts of other vitamins and minerals. It has a pleasing nutty flavor when toasted, and therefore can be used to enhance the flavor as well as the nutritive value of other foods. You might even sprinkle wheat germ on cornflakes.

□ ADDITIVES IN CEREALS □

The labels show BHA, partially hydrogenated vegetable oil, lecithin, calcium caseinate, coconut oil, BHT, tricalcium phosphate, wheat gluten, defatted wheat germ, calcium carbonate, trisodium phosphate, caramel coloring, malt syrup, malt flavoring.

Aside from the normal cereal ingredients, processed dry cereals may contain other ingredients. Many of the complicated-sounding chemical names are the names of the vitamin and mineral additives that fortify the food nutritionally. These are discussed thoroughly in chapter 3, "Nutritional Chemical Additives: The Vital 21." Once again, the vitamins are identical whether synthetic or natural. The law simply requires that when synthetic, they be labeled with their chemical name. If we labeled the natural vitamins by their chemical names, we would have to use many of the same names.

BHA and BHT prevent oxidation — they are antioxidants. As mentioned earlier, there is a considerable body of opinion that the drop in the incidence of stomach cancer in the United States may be because of the use of these two materials. There is absolutely no evidence of any harmfulness, although many disagree that they are necessary. In dry cereals, they may be used up to the level of 50 parts per million of the cereal and are classified as food additives, which are not GRAS. Based on this information, I would seek out foods with BHA and BHT in them, rather than avoid them.

Partially hydrogenated vegetable oil or coconut oil is added to

□ SUBSTANCES ADDED TO CEREALS □

Materials Added	Legal Status (and GRAS class)	Purpose
Annatto extract	Color Additive	Coloring
BHA	Additive*	Antioxidant
BHT	Additive*	Antioxidant
Calcium carbonate	GRAS (1)	Nutrition
Calcium caseinate	GRAS (1)	Nutrition and flavoring
Caramel color	GRAS (1)	Coloring
Defatted wheat germ	GRAS (food)	Nutrition
Lecithin	GRAS (1)	Emulsifier
Tricalcium phosphate	GRAS (1)	Nutrition
Trisodium phosphate	GRAS (1)	Sequestrant; nutrition
Wheat gluten	GRAS (food)	Toughens the cereal

*GRAS (3) in fats and oils, otherwise an additive.

cereals to make the product more tender. Both are saturated fats and, nutritionally, add nothing but the least desirable form of calories.

Lecithin is extracted from vegetable material, usually soybeans, and is used primarily as an emulsifier, an aid in blending. It is actually a food component, a perfectly wholesome material.

Calcium caseinate is derived from the protein of milk and is used to supplement the protein content of cereals. Calcium caseinate is a very-high-quality-protein source only to be welcomed.

Tricalcium phosphate is a source of calcium and phosphorus for nutrition, and is also used to prevent lumping of some products.

Calcium carbonate would be used to increase the calcium content of the cereal for nutritional purposes. It has the same chemical composition as chalk and is GRAS.

Trisodium phosphate is added to cereals to aid in processing so that the flakes don't break readily. While it is rarely added purely as a nutritional supplement, it does provide phosphorus along with serving the needed processing purpose.

Caramel coloring is made by heating one of the common

types of sugar, with or without added acid or alkali. To protect against contamination primarily from materials that might be included in the acid or alkali, there are very strict limits on the amount of lead, arsenic, or mercury that might be present. The law states that the amounts of acid or alkali to be used should be consistent with "good manufacturing practice." While there is no reason for alarm when this ingredient is used, in my opinion, there is no valid purpose for using caramel color in a dry cereal, and it is absent from most. It is used to modify the dull color produced in one of the very-high-sugar cereals. Caramel color is handled legally under the Color Additives Amendment of the Food, Drug and Cosmetic Act and therefore is listed as a color additive.

Artificial color: While there are some foods that would be absolutely unacceptable without artificial coloring (like maraschino cherries), it is difficult to understand why it would be necessary in a breakfast food. Apparently, the manufacturer wants to correct an unappetizing color resulting from formulation. For example, one highly vitamin-and-mineral-fortified cereal (100 percent RDA of each of the added vitamins and minerals) is artificially colored with annatto extract — a vegetable color. This is probably done to correct what is considered an off color resulting from the high iron content and a blend of three different cereal grains. Certainly the intent is noble, and the nutritional value of the cereal is excellent, but I wonder whether the manufacturer ever really established the consumers' desire or commercial necessity for adding the artificial color. Further, we are confused by the claim that the color comes from a natural source. The annatto extract is prepared by using one or more of a number of solvents such as acetone, hexane, or alkaline propylene glycol. Annatto is subject to the same regulations as caramel color.

While I can't really object to the use of these materials, every time a material is added that cannot be proven to be absolutely safe (and none can), I ask myself whether it is necessary, and if it is not, I search for a product without the artificial color.

To summarize, cereals should be an important part of our diets. If you are watching your weight, avoid the so-called natural cereals; they are higher in fat, carbohydrates, and calories, and lower in vitamins and minerals than the processed breakfast foods. Also, be wary of health claims (which are not supported on the label) sometimes made by store clerks. The law prevents false label claims, but cannot control what salespeople say in a store. I personally use the vitamin-fortified breakfast foods, with low sugar and the smallest number of ingredients — avoiding artificial flavors and colors simply because I object to their unnecessary use in this class of foods. I also use the simple oatmeal and wheat cereals as good sources of complex carbohydrates, when I want a hot, fortifying breakfast.

□ QUICK-COOKING AND PRECOOKED CEREALS □

Quick-cooking oatmeal differs from regular oatmeal in that the flake is thinner and therefore the water penetrates faster. Normally, oatmeal is made by steaming the oats and passing them between rollers to flake them. For quick cooking, the oat is chopped first, then steamed and flaked, resulting in a thinner flake, which cooks in 1 minute rather than 5 minutes.

Instant oatmeal is treated so that the water gets into it even faster, requiring only the addition of hot water and no cooking whatsoever.

Grits, that is, corn endosperm, is made quick-cooking by steaming and lightly cracking the grit so that water can penetrate more readily.

Quick-cooking or instant rice is actually precooked and dried.

All of the processes are physical. There is no chemical treatment involved: only heating, steaming, drying, chopping, flaking, toasting — all normal, household food-processing steps carried out in a food factory with no mysterious additions.

Whether you like the taste of quick-cooking products as well as the regular is another matter.

Chapter 9

PASTA

Spaghetti, macaroni, vermicelli, noodles — all are "macaroni" products, according to government standards. Contrary to what most people think, they did not originate in Italy. The story is that Marco Polo introduced the pastas to Italy when he returned from China in the thirteenth century.

Pasta products are becoming more popular in the United States, and we now consume an average of 7 pounds, dry, per person per year, which is equal to 21 pounds of cooked pasta per person per year. That may seem like a lot, but not when compared with Italy's 87 pounds per person per year dry — nearly 270 pounds cooked!

Pasta products vary tremendously in quality, even though they are standardized and must meet government requirements. The key element of quality is firmness of the pasta after cooking, as implied by the Italian term *al dente* ("to the tooth"). Limp, mushy spaghetti can result from two things: first, overcooking, the most common cause; second, the ingredients — specifically, the kind of wheat flour used.

□ BASIC INGREDIENTS □

There are a number of different types of wheat grown, and while it is possible to make pasta from all of them, only one is ideally suited: durum wheat. Durum wheat is completely different from wheat used for cake and bread. It cannot be used for baking at all, but has the desirable property of retaining its texture when made into pasta and cooked in boiling water.

When wheat of any kind is milled, the first step is to remove the outer or bran layer, which is brown, and to "purify" the endosperm, which is the white, inner part of the wheat kernel (see page 111).

With ordinary wheat, if the endosperm is cut into little chunks, the product is called farina. In other words, farina is white flour that hasn't been ground fine.

With durum wheat, these same endosperm chunks are called semolina. Semolina is simply unground durum flour and is the best material to use for pasta products. It has the best taste and texture, and it resists mushiness during cooking for that *al dente* quality.

Durum granules are virtually the same as semolina, except that durum granules can contain 20 percent of durum flour while semolina must contain 3 percent or less.

Unfortunately, there is not enough durum wheat grown to supply all the semolina needed for pasta products, so that other flours are sometimes used; for example, ordinary baking flour, either blended with semolina or used alone.

Once water is added to the flour, the pasta is formed by machine into the desired shape and dried.

Durum is rarely used for noodles. This macaroni product is made using eggs to moisten ordinary wheat flour. The dough is then rolled thin and cut into ribbons and dried.

Noodle standards permit the use of eggs in a number of forms as long as the finished product contains 5.5 percent dried egg solids, whether added as dried eggs along with water or as fresh eggs.

□ BASIC SHAPES □

The most popular pasta products are spaghetti, macaroni, vermicelli, and noodles.

Macaroni is tube shaped — it has a hole running down the center.

Spaghetti may be either a hollow tube or a solid cord.

The only thing that distinguishes one from the other is the diameter of the tube on the outside. Spaghetti, to be called that, must be more than 1/16 inch and less than 1/9 inch in diameter. Macaroni must be more than 1/9 inch and less than 0.27 (about ¼) inch in diameter.

Vermicelli is not a tube but a solid cord, and must be less than 1/16 inch in diameter.

Noodles, as mentioned earlier, are ribbon shaped.

□ LABELING □

Labeling rules as of this writing in 1981 do not require listing of all the ingredients in all standardized foods; only the optional ingredients must be declared. Only a few of the brands of pasta products actually show an ingredients statement. Most merely state claims such as "made from number one semolina" or "made from semolina and farina." Few brands contain optional ingredients.

Here are the ingredients permitted where listing is required: egg white or eggs up to 2 percent; salt; gum gluten; nutritional additives in enriched products (listed by their chemical names, such as thiamine mononitrate for vitamin B_1).

Other ingredients that are permitted and must be listed if present are: disodium phosphate, in quantities from 0.5 to 1.0 percent of the finished product (it is GRAS); glyceryl monostearate, up to 2 percent (GRAS); onions, celery, bay leaf, garlic, or any two of these seasonings.

Gum gluten, when used, minimizes soggy pasta if ordinary flour is used and in general improves the bite of the product.

Salt is used not only for flavor — in fact, not enough is added for flavor — but also to help in the processing of the product. Salt tends to toughen protein.

Egg white or eggs, in the quantity used in pasta products (not noodles), act primarily as a binder to prevent disintegration and excessive softening of the produce during cooking. This is especially true of egg whites. Egg yolks are used for color.

Glyceryl monostearate, which is permitted up to 2 percent, is sometimes used to facilitate production. Glyceryl monostearate is GRAS.

Disodium phosphate, permitted up to 1 percent, is added to make quicker-cooking pasta. It is GRAS.

That is the story on the simplest pastas of all: spaghetti, macaroni, and the other shapes.

Noodles, unlike macaroni products, require all ingredients listed on the label. I can't imagine why there is a difference in labeling requirements, but it won't be long before all ingredients will be listed on all products.

□ OTHER VARIETIES □

All other varieties mentioned are also sold as *enriched*. The enrichment level is such that 2 ounces dry of any enriched pasta will provide 35 percent of the RDA of thiamine; 15 percent each of riboflavin and niacin; and 10 percent of iron.

Enriched macaroni with fortified protein: the protein content is raised to at least 20 percent as compared with the usual 14 percent, and the protein must be of high nutritional quality. This product may also be enriched with calcium, which is not permitted for ordinary enriched pasta.

Milk macaroni products: originally products made by using milk instead of water to moisten the dough, but now any form of whole milk can be used, including concentrated, evaporated, dried milk and skim milk plus butter.

Nonfat-milk macaroni products: same idea as the one above, but using nonfat milk instead of whole milk.

Vegetable macaroni products: the most common one is green — made with spinach. The standards require that not less than 3 percent of the finished dry product be vegetable *solids.* This would be about the same as using the ground fresh vegetable instead of water to moisten the dough.

Enriched vegetable macaroni products: enrichment plus vegetables.

Whole-wheat macaroni products: made with whole-wheat durum or ordinary flour or a blend. The use of eggs, disodium phosphate, and gum gluten is *not* permitted in these products.

Wheat and soy macaroni products: same as the basic pasta products, but must contain not less than 12.5 percent soy flour in the blend used.

Similar varieties exist for noodles.

In general, the required names for these products are very descriptive and a good guide to their composition once you have read this material.

□ SUBSTANCES ADDED TO PASTA PRODUCTS □

Material Added	Legal Status (and GRAS class)	Purpose
Citric acid	GRAS (1)	Acidity adjustment
Caramel color	GRAS (1)	Coloring
Disodium phosphate	GRAS (1)	Produces quicker cooking
Glyceryl monostearate	GRAS (1)	Makes more tender
Gum gluten	GRAS (food)	Firms up the pasta
Modified food starch	Additive (1–5)*	Thickens sauce
Monosodium glutamate (MSG)	GRAS (2)	Flavor enhancer

*Modified food starches are not now specifically identified on the label. They may be anything from class 1 to 5, depending on which starch is used.

□PASTA PRODUCTS□

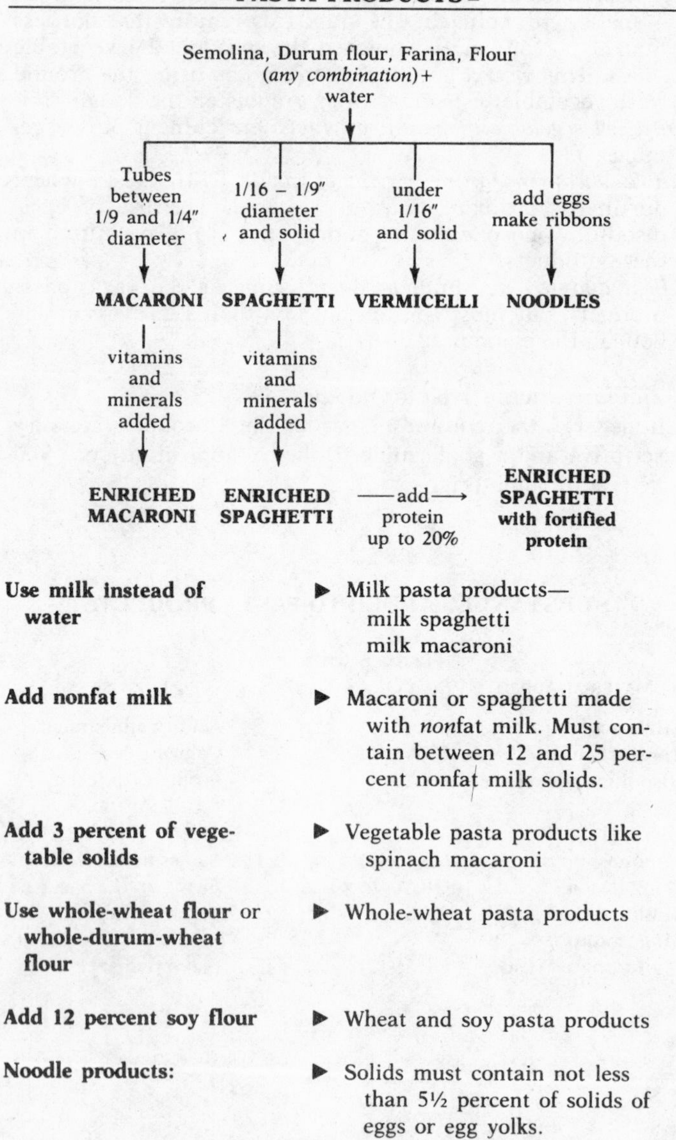

Noodle products: ▶ Solids must contain not less than 5½ percent of solids of eggs or egg yolks.

□ NUTRITION AND PASTA □

As the government has been telling us, we need to eat less sugar and more complex carbohydrates. Pasta is one of the best sources of those complex carbohydrates and, for the money, packs a real nutritional wallop.

Even noodles with egg added provide only 3 grams of fat in a portion; about 12 percent of the calories are fat calories. In the case of spaghetti, only 1 gram of fat is present.

It is no wonder that so many people have learned to make a main meal entree based on pasta. Just stop to consider: 4 ounces of spaghetti (which would cook up into 12 ounces), plus ¼ pound of ground beef in some tomato sauce, with a 2-ounce serving of green beans on the side, would provide:

Protein: 30 grams, nearly half the RDA
Calories: 610–700, only one-third of the daily average
Thiamine: 70% of the RDA in the spaghetti alone; 10% from the beef; 3% from the beans
Riboflavin: 30% from the spaghetti; 10% from the beef; 3% from the beans
Niacin: 30% from the spaghetti; 25% from the beef; 1% from the beans
Iron: 20% from the spaghetti; 30% from the ground beef

This is a truly well-balanced meal, at a really low cost. The spaghetti would cost less than 25¢, and the entire main dish would probably be less than $1.50, for one-third of the day's calorie requirement, 83 percent of the thiamine, 43 percent of the riboflavin, and 56 percent of the niacin.

Pasta products should be an important part of our diets for their low-fat content (if we don't add too much in the sauce), complex carbohydrates, and their wonderful ability to serve as a base for the higher-protein foods that can be combined with them: chicken, fish, cheese, and beef.

□ CANNED PASTA □

A few words should be said about canned pasta, which includes macaroni and cheese, spaghetti, and cheese ravioli. The ingredients statements on these products are more complicated for two reasons: first, since these foods are not standardized, all ingredients must be declared; and second, the canning process and economic considerations introduce some new materials.

The simplest of them introduces only citric acid as an added ingredient, the other ingredients being foods to make the sauce. Citric acid (GRAS) enhances the acidity of the tomato sauce. Hydrogenated oils, caramel color, modified food starch, and monosodium glutamate are the only other materials not commonly used in homemade sauce. Of these, the modified food starch and the caramel color are additives.

Nutritionally, it is difficult to compare the label of canned, ready-to-eat pasta, sauce and all, with homemade pasta and sauce. You are simply unable to do so without full information on the ingredients used in the sauce. Two leading brands of sauce do not state their nutritional analysis.

What is clear is that the canned spaghetti uses less dry spaghetti to make a portion than the 2 ounces shown as the serving size on the dry spaghetti label. We can tell this because the spaghetti alone, 2 ounces of it, no matter which brand, will supply 220 calories. The serving size of one canned spaghetti with sauce (7⅜ ounces) supplies only 160 calories.

This is important only if cost concerns you: you're paying for more water and less food.

Chapter 10

MEAT, POULTRY, AND FISH

These foods are the mainstay of our diets. They supply high-quality protein as well as the important element of satisfaction at so many meals—especially the main meal of the day. In our culture, we feel that a main meal is incomplete without one of these foods. We spend 26 percent of our food money on this food group.

Depending on our weight, the recommended daily allowance for protein runs from 45 to 65 grams per day, with an extra 20 grams for pregnant or lactating women.

Protein is one of the absolute essentials of our diets. We can manage very well with restricted consumption of fats or carbohydrates, but not with insufficient protein. This is the food element needed to rebuild body tissue, to supply the essential elements for growth of muscle tissue.

Not all protein is of equal value to the body. Gram for gram, meat protein is almost twice as valuable as wheat protein. This means that a person would have to eat twice as much wheat protein as meat protein to get the same body-building effect, which means far greater calorie intake for each gram of protein gained. To obtain 65 grams of protein per day from cereals, we would need to consume about 650 grams per day of the dry cereal (flour, for example). To consume 650 grams of flour, we would need to eat about 1300 grams of bread (made without milk or sugar in this example). That's nearly 3 pounds of bread. Along with the protein, we would consume over 2000 calories — and would still be a long way from meeting our total daily vitamin and mineral requirements. Thus, it is not very practical for us to get our daily protein requirement from cereal grains alone, although the addition of legumes (beans, peas, soybeans), plus eggs and milk, makes a healthy vegetarian diet possible.

While it is important for us to be informed of the nutritional composition of this food group, it is quite difficult because most of our consumption is of unpackaged, fresh, and unlabeled products, except for fish. Only about a third of the fish we buy is unpackaged and fresh, with the balance being packaged, frozen, or canned, and therefore labeled.

The chart at the end of this chapter shows the nutritional analyses of these foods. In the case of fresh food, the analysis is shown for a quarter of a pound — a fairly normal portion. Don't be overwhelmed by all this information. There is a simple set of facts to remember that will guide you well.

Fact 1 (protein content): No matter whether it's beef, pork, lamb, chicken, or fish, a quarter of a pound of the food will give you *about* 20 grams of protein, that is, about a third of the recommended daily allowance. What varies from food to food and from cut to cut is water content, fat content, and the number of calories.

Fact 2 (caloric value): A quarter pound of most fresh *fish* yields about 100 calories. *Lean beef,* carefully trimmed, provides 200 calories; *chicken,* 150 calories; *untrimmed beef,* most cuts, 400 calories; *lamb,* untrimmed, 350–400 calories; *pork,* 350 calories; *hamburger,* average fat, 300 calories. For more exact figures, refer to the chart.

Fact 3 (cholesterol content per quarter pound): Whether beef, veal, lamb, or pork, a quarter of a pound will supply about 80 milligrams of cholesterol. Not much different from fish and chicken — not even worth trying to remember the slight difference.

However, when we get to the organ meats and shellfish, this story changes. The cholesterol provided per quarter pound of these is:

Liver: 340 milligrams
Shrimp: 140 milligrams
Lobster: 230 milligrams
Crab: 140 milligrams
Brains: more than 2000 milligrams
Sweetbreads: 280 milligrams

We could find no information on cholesterol in frankfurters and sausages, but it should be assumed to be high because of the use of organ meats in making them.

□ INCIDENTAL ADDITIVES □

In order to raise the millions of animals required for our food supply, many pitfalls must be overcome or avoided. First of all, in order to feed them, we must raise enormous quantities of grain, which in turn means the use of fertilizers, weed killers, pesticides, and fumigants to be sure there will be a sufficient crop. Second, animals are subject to many diseases, and in order to be sure that we are supplied with healthy animals, all sorts of drugs are used, especially antibiotics. It is safe to say, I think, that without the use of these substances, at least some of them, we simply would not have an adequate meat or poultry supply, or if we did, the prices would be beyond the reach of most people.

This is a far cry from the early twentieth century when all animals, regardless of condition, were led to slaughter, as exposed in Upton Sinclair's book *The Jungle*. This book led to the passage of legislation that began to protect our meat supply, and a system of inspection that assures us of safe meat and poultry.

It is the U.S. Department of Agriculture (USDA) that is responsible for the regulation and control of the safety, wholesomeness, and labeling of meat and poultry. Every slaughterhouse facility is inspected by a government inspector, as is every processing plant.

Where regulations under the Wholesome Meat Act or the

Wholesome Poultry Act do not apply, FDA regulations take over, as in the case of labeled, packaged meats.

The USDA checks for more than 100 substances that might be present in the meat as a result of the many steps required to prepare it for sale. New methods of analysis are extremely rapid and sensitive; substances can be detected in amounts of 1 part per billion and less. Meat is tested for pesticides; heavy metals, such as mercury; herbicides; many drugs, especially antibiotics and sulfa drugs; and hormones. As recently as August 1979, a new test for antibiotics was developed that gives results in hours, rather than days, and is now in use. The amounts permitted in tissue run from zero to a few parts per billion.

In the past, concern was expressed especially about two substances in our meat and poultry supply: penicillin and diethylstilbestrol (DES), a hormone. At this time, the use of DES is totally prohibited. But, even when it was permitted, no residue was allowed, ever. With penicillin, only one-half part in a billion is permitted in beef and none in pork.

Neither of these amounts of penicillin or stilbestrol is sufficient to have any effect on a human being in themselves. However, there are three factors that must be considered that are the basis for concern by those opposed to the use of these substances.

First: *Can these levels be maintained?* Millions of animals are raised on thousands of farms by hundreds of thousands of people. Granting the best intentions in the world, mistakes will be made, and since the FDA and USDA can sample only a minuscule portion, it is inevitable that some material that exceeds the limits will reach the consumer.

Second: *Might these substances have other effects that could indirectly affect people?* For example, could diethylstilbestrol have adverse effects that we don't know about yet?

In the case of widespread use of penicillin in cattle, swine, and poultry, is it possible that this practice contributes to the development of penicillin-resistant bacteria, which will make it more difficult to treat human diseases in the future?

Third: *Should these substances ever have been permitted before there were sufficiently accurate and rapid methods of testing?*

Both substances were permitted for years before there were sufficiently sensitive and speedy tests. (Canada, incidentally, has never permitted the use of stilbestrol.)

The reason for the elimination of stilbestrol was that when it was administered to women as a drug, it seemed to increase the incidence of cancer. True, the dosage was thousands of times greater than that normally used on cattle, and one should not conclude that there was any mass health hazard in meat. But why was it used on cattle in the first place? To fatten animals before slaughter and to increase the yield! A gain of that kind, reflecting only a conceivably slight price advantage, is simply not sufficient to warrant any risk at all to people.

For your interest, stilbestrol was given to animals in two ways, either in their feed, where it could be pretty well controlled, or by implantation of a pill in the ear skin. The rules called for the elimination of the drug at least two weeks before slaughter so that there would be none left in the meat. Well, it's easy to stop using the stilbestrol in the food, but how do you make sure you have removed the pill from behind the ears of four thousand unruly animals? Seems to me you're bound to miss a few ears!

Penicillin and other antibiotics are very commonly used to prevent disease in almost all chickens, cattle, and pigs raised. There is sufficient concern about the problem of residue and, more important, the development of disease-resistant strains of bacteria that can affect man, to prohibit the use of antibiotics in animals in many countries, except to cure disease. Also, an FDA task force on antibiotics recommended that the same antibiotic should preferably not be used for both human and animal applications. This recommendation was based on results of studies that showed that resistant strains did develop. New antibiotics have been developed for use on animals only. These have been approved in the United States and other countries.

While it is not possible for us to know exactly what is in the meat we buy, so that we cannot make a choice based on knowledge, the likelihood of any significant hazard appears to be infinitesimal.

The risk of incidental additives that might exist in our

present system and practice are preferable, I believe, to the alternatives: the hazard of eating contaminated food from diseased animals, a far more serious possibility, or the danger of our feed crops' destruction by pests and weeds, which would drastically reduce our supply of meat.

There is not a great deal we can do about incidental additives as shoppers, but we can and should support consumer groups that attempt to influence legislators to prevent practices of questionable value.

As individuals, we can exert some control by not eating too much of any one kind of meat, especially liver, the organ in which pesticides, hormones, and many other incidental additives accumulate if they should be present. In addition, liver is higher in cholesterol — more than four times higher than beef muscle.

■ FISH

The mercury scare of the early 1970s didn't turn out to be much, but it did start a better program of testing fish for heavy metals. We can assume that the fresh fish we buy is free of contaminants.

□ PROCESSED MEATS □

About one-third of all the meat consumed in the United States is not fresh meat, but has been processed in some manner. Hams, sausages, salami, bologna, liverwurst, hot dogs, and bacon are all processed.

Since time immemorial, people have preserved meat by smoking it and by salting. Not only common salt has been used but also saltpeter — sodium nitrate. It was discovered that sodium-nitrate treatment of meat, fish, and poultry prevents the growth of botulinus bacteria, which is responsible for the potent killer, botulism (see chapter 2). In addition to preventing botulism, sodium nitrate also produces what we have come to consider the "natural" color of hot dogs, bacon, ham — a pink or red color, rather than gray, which is the true natural color of these products when cooked.

The original method of curing (treating with salt and sodium nitrate) consisted of burying the ham in a mixture of dry salt and sodium nitrate for a number of weeks, then washing the salt off and hanging the ham in the smokehouse for a day or so, so that the heat of the smokehouse could raise the temperature of the ham or bacon high enough to kill most of the bacteria in it, producing a ham or slab of bacon that could keep safely at room temperature for weeks. All kinds of wood, and even corn cobs, were used to produce the smoke, with hickory being most popular.

Another old-fashioned method of curing is to make a solution of sodium nitrate and seasonings, including salt, and soak the meat in it for up to six weeks until the cure solution has penetrated the meat to preserve it and change the color. Afterward, the meat is rinsed or washed and then smoked as described before.

Because these methods are so time-consuming, shortcuts were found, such as injecting the cure solution into the arteries of the slaughtered animal or injecting curing solution in a number of places in the meat, and then smoking after twenty-four hours or less.

Two factors have contributed to the decline in popularity of smoking meat. First is the worry that some components of the smoke are carcinogenic. Smoking has now been replaced in most large plants by spraying the meat with smoke flavor, which is produced by condensing the smoke-flavor element without the components that are believed to be carcinogenic materials. The other factor is a developing concern in large cities about the environmental effects of smokehouses and discharged smoke.

In 1910, it was discovered that the sodium nitrate was converted by bacteria in the curing solution to sodium nitrite, and that it is sodium nitrite that actually does the work in the meat. Consequently, processors started to use sodium nitrite along with, or as a replacement for, the nitrate. By blending the two, manufacturers of cured meat can regulate the speed at which curing takes place.

One of the problems with the fast-injection cure process was getting the meat to absorb the liquid that was injected into it. Because of this, sodium phosphates were included in the cure solution. This in turn caused the ham to gain weight in the form of water—lots of it—so much that regulations were required to limit the amount of gain to 10 percent or less.

When hams have been treated in this way, their label must say "water added," and that doesn't mean added to a pot, it means into the ham itself, so that when you buy it by the pound, you get nine-tenths ham and one-tenth water for every pound.

Hams that have been processed in the old-fashioned way *lose* 10 percent water, so that the difference between the old type and new is 20 percent water. This difference is reflected in the taste of the ham. The same thing is true of bacon but is less significant because so much of the bacon is fat, and the water cannot get into the fat.

Hot dogs, salami, and other sausage products have the sodium nitrate and nitrite added either in the curing process or directly to the sausage mixture before it is cased. This is done for preservation and, in the case of hot dogs, for attractive color as well.

Corned beef is simply beef that has been cured but not smoked. Smoked fish are dry cured: placed in a bed of salt and nitrate and/or nitrite and smoked afterward.

All of this curing and processing, other than the smoking, is done at rather low temperatures: 30° to 40° F.

Not all sausages are cured, but those that are not must be kept frozen until cooked.

Major changes in the amount of nitrite permitted, the use of vitamin C derivatives, and the testing program for the presence of nitrosamines in bacon have greatly reduced any hazard that may have existed (see "Food Additives and Cancer" in chapter 1). But although the hazards have been reduced, we are still concerned about the links between nitrites and smoked foods and cancer. What are we supposed to do? Before we panic and deny ourselves all the pleasures of hot dogs, bacon, and ham, there are some other things to know.

Only about 20 percent of our nitrate consumption comes from such foods! We are continually ingesting nitrate in water and in vegetables, which our saliva converts partly into nitrite. There is some opinion that the natural sources of nitrite are far more significant in contributing to cancer than the nitrites in processed foods. So it appears that even if we did cut out all such foods, we would not solve the problem.

We don't outlaw the use of nitrates and nitrites because millions of people habitually store the foods containing them at room temperature. Without nitrates, these foods would not be preserved, yet there is a great risk that people would still handle the products as cured meat, out of pure habit. Therefore, people responsible for public health believe that there would be far more deaths from botulism than from any conceivable number of cancer cases that might develop from the use of nitrates. If we could be sure that cured and smoked meats would all be stored at freezer temperatures, we could safely remove all nitrates and nitrites immediately. The only problem we would then face would be less attractively colored bacon, ham, and hot dogs — they would be gray, rather than pink or red.

There are two things about cured foods that make them so popular:

1. They taste good.
2. They are extremely convenient.

Because of these advantages, there are many people who eat cured foods many times each *day,* every day. Bacon for breakfast, liverwurst sandwich for lunch, maybe some ham at night. Certainly, when you stop to think about foods conveniently available for sandwiches, cured meats lead the list.

My personal conclusion is to go easy and to try to limit consumption of nitrite-cured foods to once or twice a week. It is not easy, and it is about time some food manufacturers began to develop some tasty and convenient replacements.

There is another consideration. Many of these foods by and large are high in fat, low in protein, and, in the case of liverwurst, high in cholesterol. For these reasons alone, their consumption should be moderate. Some hot dogs today contain less than half the protein content of beef—nearly the same amount of protein as the bread or roll with which they are eaten!

■ CURED MEAT LABELS

There are some ingredients listed on meat labels that are fairly universal, such as water, sugar, sodium nitrate, sodium nitrite, sodium erythorbate, ascorbic acid, sodium ascorbate, and salt.

A good way to determine percentage of some ingredients is to look for salt on the label first. Most bakery products contain 1 percent salt. Savory products, such as cured meats, contain 3 percent salt, and truly salty products, like pickles, may contain as much as 5 percent salt or more. We can thus judge the quantities of other ingredients by where they are listed in relation to salt.

We've discussed nitrates and nitrites. The sodium erythorbate, sodium ascorbate, and ascorbic acid are all vitamin-C-related, and as we've mentioned, they produce better color with less nitrate and may be useful in preventing the formation of carcinogens. Very small amounts of sugar are used, primarily as part of the seasoning, and can be ignored as a significant source of sugar.

Here, then, are some representative labels for common cured meats.

Canned ham: "Water, salt, dextrose, sodium phosphates, sodium ascorbate, sodium nitrite. With natural juices —gelatin added." The dextrose is corn sugar, much less than 3 percent (since it is behind salt on the label). The sodium phosphates—GRAS—are added to keep the water in the ham; the gelatin, to help mold the ham and retain its shape when removed from the can and sliced. This ham is not smoked.

Bacon: "Water, salt, sugar, sodium phosphate, sodium erythorbate, sodium nitrite." The sugar content must be very low: under 3 percent. It is a good idea to use bacon that has been cured with sugar or dextrose. It will brown faster than unsugared bacon. The carcinogens receiving the most publicity and causing the greatest concern have been the nitrosamines formed when bacon is fried. If these undesirable substances are formed at all, they are formed at high temperatures; the quicker-browning bacon will brown at lower temperatures, thus providing some protection. (It might be a good idea for bacon producers to cure bacon so that it would be burned and unpalatable at the dangerous temperatures, to provide a built-in safety system.)

Another brand proudly announces: "No Sugar Added." Remember, the amount of sugar is negligible. In four strips there might be ¼ teaspoonful at the most.

Ascorbic acid has been found to inhibit the formation of nitrosamines, and current regulations *require* the use of ascorbic acid, sodium ascorbate, or sodium erythorbate.

Finally, as a precautionary step, bacon is now monitored for nitrosamine formation by frying for three minutes on each side at 340° F and then testing for nitrosamines.

In actual practice, the amount of sodium nitrite present in bacon is considerably less than 100 parts per million, and in some cases less than 20 parts per million.

I think the government has done an outstanding job in investigating a food preservation practice nearly 2000 years old, learning of a long-shot danger, and acting to minimize it.

Sausages: Compare these three ingredient statements.

1. "Pork, salt, spices."

That's as simple as you can get — but it is for a frozen pork sausage.

2. "Pork, partially defatted pork fatty tissue, water, salt, spices, sugar."

This is also frozen. I think the partially defatted pork fatty tissue means slabs of pork fat with some of the fat cut away.

3. "BHT, BHA, with citric acid added to protect flavor. Fully cooked. Ingredients: Beef, water, salt, sugar, spices, dextrose, monosodium glutamate."

This is for brown-and-serve sausage, found in the refrigerator case. There is also another variety with beef and pork. The brown-and-serve product is frozen until it gets to the store and then may be kept in the freezer or refrigerator. Because it has been fully cooked, it is reasonably sterile, and thus can tolerate some refrigerated storage rather than requiring frozen storage.

The BHA, BHT, and citric acid are added to prevent fat rancidity, which could develop even in the frozen state after precooking. BHA and BHT are used in minute quantities. Our stomach cancer rate is now among the lowest in the world; as mentioned earlier, there is a theory that BHA and BHT may be responsible for the decline. Citric acid is GRAS and is a normal component of human body tissue and blood.

Liverwurst (sometimes called braunschweiger): "Pork livers, pork, bacon, salt, spices, dehydrated onions, water, sugar, sodium nitrite, and sodium ascorbate." Since liverwurst is sold and held at refrigerator temperatures, or even held at room temperature in a sandwich, it must contain sodium nitrite for safety. The water is used to adjust the consistency, and sugar at a level of less than 2 percent is used for taste blending.

Bologna: "Pork, water, beef, corn syrup, salt, flavoring, dextrose, paprika, ascorbic acid, sodium nitrite." No new ingredients on this one. But note that beef follows water on the label; my guess would be that it contains less than 5 percent beef. The paprika is present primarily for color, since it doesn't contribute much flavor in small quantities.

Salami: "Pork, water, corn syrup, salt, flavoring, dextrose, monosodium glutamate, paprika, ascorbic acid, sodium nitrate." Monosodium glutamate is used to enhance the flavor; it is not in itself a flavoring material. It is universally used in the Orient for this purpose. It is GRAS, class 2. In fact, every time we eat bread, we convert part of the protein into glutamic acid and absorb it as sodium glutamate.

There is not much difference between bologna and salami

except for the texture of the meat, which is coarser in salami and smoother in bologna. Both are overgrown sausages that may be sliced.

Pepperoni and other sausages will be found to contain the same ingredients, differing in the kind and amount of spices used.

Hot dogs, also called franks or wieners, are the most popular products of the sausage group. The meat sources are beef, pork, or, recently, chicken, or a combination of them. Beef franks contain: "Beef, water, corn syrup, salt, flavoring, dextrose, paprika, sodium nitrate, sodium ascorbate." The flavoring may vary, and the sodium ascorbate may be replaced by sodium erythorbate.

What really is important to know is the nutritional analysis, and only a few brands now show it. One brand shows 13 grams of fat and only 5 grams of protein in one hot dog. If the meat were all lean beef, the amounts of fat and protein would be equal. (Chicken hot dogs contain one-third less fat.)

When you can, shop for hot dogs by their protein content — the higher, the better — and reject those with higher fat than protein.

The source of meat in hot dogs is any part of the carcass: hearts, intestines, anything. That's why hot dogs are relatively cheap and high in fat. They taste good because they are spiced and salted and smoked, or smoke flavored.

Eat hot dogs for enjoyment, not for significant nutrition, now that they have evolved into a low-protein, high-fat food.

□ CANNED MEAT SPECIALTIES: TWO EXAMPLES □

Chili: "Water, beef, beef heart meat, flavorings, wheat flour (bleached wheat flour, malted barley flour, potassium bromate), tomato paste, textured vegetable protein (soy flour, caramel coloring), modified cornstarch, dehydrated onions, cereals, salt, soy protein isolate, paprika, hydrolyzed plant protein."

Where's the chili? We're not told, unless it is in the "flavorings." This particular label demonstrates a number of instances of illegality and carelessness. The cereals are not identified, which is required by law. The flavorings, if they are natural and also include spices, would look more appealing on the label if they were identified as such. The use of bleached wheat flour, with the additives in it, which are needed for efficient bread production only, is pure laziness. There is plenty of unbleached wheat flour with no additives available for this purpose, which is merely to adjust consistency.

Explanation is in order for "textured vegetable protein." This is not a food additive but actually soybean flour that has been mechanically and heat treated to make a food that is of major importance to us as a source of high-quality protein. Soy flour alone is rather limited in its food use because of its texture and flavor, and the texturizing process has overcome this deficiency. In this product it is not nutritionally inferior; in fact, it provides excellent nutrition at a lower cost than meat.

Hydrolyzed plant protein is an ingredient you will find in a number of savory foods. It, too, is a food, not an additive. It is produced by treating a protein, such as gluten from flour or from other plants, with acid or enzyme to split the protein into amino acids, which gives it a meaty taste, and also forms some sodium glutamate, which enhances that taste. It is totally wholesome. Soy sauce is one form of hydrolyzed plant protein —produced by enzymes—and the proteins actually used are wheat and soy, combined.

Corned-beef hash: "Water, beef, cooked corned beef (cured with salt, sugar, and sodium nitrate), dehydrated potatoes, salt, sugar, flavorings, spice, sodium nitrate."

A second brand lists virtually the same ingredients, except it lists sodium nitrite instead of nitrate.

Unfortunately, we have no way of knowing how much corned beef we get in the product. There is no nutritional analysis. In my opinion, this is another food to be eaten for fun and not nutrition. My guess is that it is loaded with carbohydrate and fat, with not much protein.

□ CANNED FISH □

Canned tuna has become a fairly important main-meal protein source, and deservedly so. The nutritional label of a seven-ounce can of water-packed tuna shows that it provides more than 100 percent of the recommended daily allowance of protein — with only 240 calories. It would take more than a half pound of most meats to supply that amount of protein, with a lot more calories.

The oil-packed product reads: "Light tuna, soybean oil, vegetable broth, and salt."

The ingredients for one water-packed brand read: "Albacore in water, seasoned with vegetable broth and salt, with added pyrophosphate." Another water-packed brand reads: "Light tuna, spring water, salt." Pyrophosphate is added to firm up texture, which obviously is not necessary since this alternate brand doesn't contain it.

Meat, poultry, and fish are the main sources of protein in the American diet. Your choices within this food group can determine whether you will consume, along with that protein, excessive fat or cholesterol; whether you will ingest large, small, or no quantities of nitrate and nitrite. The choices are great, and it is possible to select whatever nutritional combination you favor, using the chart in this chapter and the information on the labels.

Try to eat foods containing nitrite only once or twice a week. And by all means avoid having bacon and eggs for breakfast, hot dogs for lunch, followed by ham for dinner. That is just too much of the same kind of nitrite-containing, smoked, high-fat food.

Favor the low-fat meats; eat lots of fish and chicken. When eating chicken, avoid eating the skin (or feel guilty when doing so). Chicken skin has nearly twice the calories of chicken meat because of its fat content. Frying chicken doubles the calories and triples the fat! If you can be happy eating mostly veal, poultry, and fish, you can really eliminate major calorie and fat and cholesterol sources from your diet. However, if you replace the high-fat meats with high-fat dairy products (cheese), you won't have gained a thing.

Don't hesitate to write the manufacturer asking for the nutritional analysis of any of the packaged, processed products. Just say, "Please send me the nutritional analysis for __________."

□ NUTRITIONAL COMPOSITION OF MEAT, POULTRY, AND FISH □

(¼ Pound Raw Unless Otherwise Indicated)

	CALORIES	PROTEIN (grams)	FAT (grams)
Beef			
Chuck	290	21	22[1]
Steaks			
Porterhouse	436	17	40[1]
T-bone	440	16	41[1]
Sirloin	370	18	32[1]
Round	220	22	14
Hamburger			
Lean	200	23	11
Regular	300	20	23
Rib Roast	400	17	39[1]
Corned Beef (Cooked)	416	26	34
Liver (Beef or Calf's)	160	22	5
Kidneys (Beef, Pork, or Lamb)	145	18	7
Veal			
Loin	204	21	12
Rib	240	21	15
Lamb			
Leg	250	20	18[2]
Loin	340	18	28[2]
Rib Chops	380	17	34[2]
Shoulder Chops	315	17	27[2]
Kidney *see beef*			

[1] About ⅔ of this fat can be trimmed, thus cutting fat to about 13 grams and reducing calories by half.
[2] Two-thirds of this fat can be removed by trimming, reducing calories by half.

	Calories	Protein (grams)	Fat (grams)
Pork			
Ham (Fresh)	340	18	29[2]
Ham (Canned)	216	20	13
Loin	340	19	28[3]
Ham (Cured)	315	20	26[2]
Spareribs	400	16	37
Sausage (Raw)	557	10	57[4]
Kidneys *see beef*			
Cured Meats			
Liverwurst	340	18	28
Bacon (Fried and Drained, 15 g.)	90	5	8
Bologna	340	13	30
Salami	340	19	29
Frankfurters (2½ Franks)	351	10	32
Poultry			
Chicken and Turkey	151	20	6
Squab	320[5]	20	26
Fish			
Cod	87	20	0
Fishsticks (Frozen; also contain 7 g. carbohydrate)	197	18	10
Flat Fish (Flounder, Sole)	88	18	1
Bass	110	21	3
Mackerel (Atlantic)	213	21	13
Porgy	125	21	4
Pollock	110	22	4
Salmon	243	25	15
Sardines (Canned and Drained)	227	27	27
Bluefish	131	22	4
Tuna (7-ounce can, water packed)	240	50	6
(7-ounce can, oil packed)	520	50	34
(4-ounce can, water packed)	137	28.5	3.5

[3]Half the fat can be removed by trimming, reducing calories to 340.
[4]About 30% lost in cooking.
[5]Higher caloric content owing to higher ratio of skin to meat.

	Calories	Protein (grams)	Fat (grams)
Tuna (cont.)			
(4-ounce can, oil packed)	297	28.5	19
(4-ounce can, oil packed, drained)	220	28.5	10
Shellfish[6]			
Clams	55	7	0
Crab	110	18	3
Shrimp	101	20	1
Lobster	101	19	2
Scallops	90	17	0

[6]Carbohydrate content very low—can be ignored.

□ SUBSTANCES ADDED TO PROCESSED MEAT, POULTRY, AND FISH □

Material Added	Legal Status (and GRAS class)	Purpose
Ascorbic acid	GRAS (1)	Curing
BHA	GRAS (3)*	Antioxidant
BHT	GRAS (3)*	Antioxidant
Caramel color	GRAS (1)	Coloring
Gelatin	GRAS (food)	Gels ham
Modified cornstarch	Additive (1–5)†	Thickener
Monosodium glutamate (MSG)	GRAS (2)	Flavor enhancer
Paprika	GRAS (food)	Seasoning
Potassium bromate	Additive	Processing of flour; no purpose in the chili
Sodium ascorbate	GRAS (1)	Curing
Sodium erythorbate	GRAS (1)	Curing
Sodium nitrate	Additive	Preservative (curing agent)
Sodium nitrite	Additive	Preservative (curing agent)
Sodium phosphate	GRAS (1)	Firms ham with water in it
Sodium pyrophosphate	GRAS (1)	Firms fish and holds water in it

*GRAS (2) in fats and oils. Otherwise an additive.
†Modified food starches are not now specifically identified on the label. They may be anything from class 1 to 5, depending on which starch is used.

Chapter 11

FROZEN FOODS

Pizza, Entrees, Vegetables

□ PIZZA □

Immensely popular pizza is no empty-calorie snack. It is a highly nutritious, inexpensive food that can supply up to 25 percent of the day's total requirements of high-quality protein, calcium, and a number of vitamins.

For frozen pizza alone, American consumers spent nearly a billion dollars in 1979!

What is pizza? Originally, it was a simple tomato and cheese pie, modified on occasion with the addition of sausage or anchovies. Pizza dough, used for the crust, consisted originally of wheat flour, water, yeast, perhaps a bit of salt. The pie filling consisted of tomatoes, tomato puree or sauce, cheese, a bit of olive oil, perhaps some green peppers, spices, and whatever additional foods the pizza baker selected to make the simple pie more interesting.

This is still the kind of pizza you get in an old-fashioned pizza bakery. The crust is very chewy; indeed, because the surface is so crisp and the interior so chewy, it can be cut only with a pastry wheel, not an ordinary knife, and really can be eaten only by curling up the cut wedge and biting into it. Alas, this kind of pizza is not to be found in the frozen-food cabinet. The reasons are many — some good, some not so good.

In many ways, a modern frozen-pizza factory is like an automobile-assembly plant. Each of the ingredients must be prepared in advance, including the unbaked crust (sometimes the crust is even baked in advance), so that the finished pizzas can come marching down the conveyor belt ready to go into packages and be frozen at a very high speed — more than one every second.

The components prepared in advance must be preserved in some way so that they don't spoil before being put into the pizza. This leads to the use of a variety of ingredients and additives that would otherwise be unnecessary. Since everything must be done at high speeds to make the product "economical," other additives must be used to make it possible for depositors, dough shapers, filling machines, cheese shakers, and so forth, to work properly.

And finally, since the product is to be frozen and shipped and will inevitably be mistreated many times along the way, ingredients must be added to try to keep the pizza texture approximately right.

We'll discuss those additives shortly, but we still have one more complication to cover. Efforts are made to appeal to everyone with pizza. Many people don't like tough, crisp crusts, so some brands include shortening in the crust to make it tender. In the interest of convenience, smaller units like "French-bread pizza" are offered. Each of these variations stimulates changes in ingredients.

One unfortunate thing about pizza is that you can't really tell if it's any good to eat by reading the label. So much depends on the handling of the dough itself, as well as the ingredients present. One beautifully simple label housed a product that had a crust like soft, factory-made white bread! The crust simply wasn't mixed, fermented, and baked right and was expanded by too much air to a spongelike texture. You will just have to try a number of pizzas to find the ones you like. They vary tremendously.

■ THE LABEL
Ingredients of Pizza

I counted sixty-two ingredients on one pizza label! And eleven on another! Nineteen on another, thirty-six on still another. Here are some of the ingredients that would cause Neapolitan pizza makers to cringe:

Potassium bromate: a dough improver. This makes dough "machine" better and enables it to puff up more when baked. Potassium bromate is a food additive that contributes nothing to taste or nutrition. It aids only in production efficiency — maybe, since not all pizza producers use it.

Sodium aluminum phosphate: a leavening agent. The pizza containing this substance did not contain yeast. Purely and simply a production convenience.

Sodium citrate: used to control acidity. Acidity not only affects taste but consistency as well. Sodium citrate is GRAS and probably useful in helping to assure a uniform product.

Mono- and diglycerides: emulsifiers that make the crust more tender. They are GRAS.

Oleoresin paprika: this sounds good, but is merely a coloring agent used in sausages. It contributes no spice value.

Modified food starch: controls consistency to prevent excessive runniness of the sauce. It is probably an ingredient in the canned tomato sauce used. It is an additive, not GRAS.

Dextrin: a food material prepared from starch. When starch is changed to sugar by enzymes, it is first changed to dextrin on the way. By stopping the action, dextrin can be made from the starch. It regulates consistency as starch does, but with a less pasty character.

Sodium stearoyl-2-lactylate: another of the ubiquitous dough conditioners the manufacturer uses to improve production. It is an additive.

Sodium metabisulfite: still another dough modifier to aid production. An additive.

Potassium sorbate: stops or slows down mold growth, probably in the cheese substitute, which must be stored before use. It is GRAS.

Textured vegetable protein: merely heat-processed soybean flour that helps to supplement meat. Nutritionally sound, a real food, with a terrible name.

Cheese substitute: contains water, casein, partially hydrogenated soybean oil, sodium citrate, potassium sorbate, lecithin, vitamin premix, sodium aluminum phosphate. Why use it? To reduce the cost. The sodium aluminum phosphate is to make the cheese substitute rigid, the casein is the milk protein out of which real cheese is made.

Xanthan gum: a material, produced by bacteria, that regulates consistency. It has characteristics that are different from any of the other agents that do a similar job, such as the vegetable gums, and is gaining widespread use in many foods. It has a very unpleasant taste, which can be detected if too much is used, but it does control consistency over a wide range of temperature and moisture conditions. It is not GRAS.

Artificial color: in a pizza? No excuse whatever.

■ THE LABEL
Nutritional Value of Pizza

Pay close attention to pizza nutritional labels. Watch for the portion size. There ought to be a legal requirement that food analyses be stated for a standard serving weight for each food so we don't need a computer to determine comparative nutritional values. There will be such a requirement in the future, as was announced by the FDA in its 1980 policy statement.

Compare the four brands in the chart below. Nutritionally, which is a better choice?

BRAND	**1**	**2**	**3**	**4**
Calories	390	229	330	240
Protein (grams)	23	8	15	11
Thiamine (% RDA)	20	13	8	25
Serving Size (ounces)	7	4	4.75	4

You really can't tell. In order to make a comparison easily, all the figures must be changed to show the nutrients *per ounce,* as I have done in the chart below.

BRAND	1	2	3	4
Calories	56	57	69	60
Protein (grams)	3.3	2.0	3.2	2.75
Thiamine (% RDA)	2.9	3.3	1.7	6.25

As you can see, Brand 1 appears to have the highest number of calories per serving, but actually contains the fewest number of calories per ounce of any of the four brands. When comparing brands, do not ignore the portion size on which the nutritional information is based.

I haven't shown all the nutrients on the label, to keep our comparison simple, but based on the information that is shown, I'd choose Brand 4 or Brand 1. Brand 3 is higher in calories and lower in thiamine, and Brand 2 is much lower in protein.

Pizza is generally high in nutritional value. Vitamin content will vary based on whether enriched flour is used or not, so look for that on the label if you want vitamins in pizza.

The other thing to look for is fat content, which will run from 7 to 10 percent of the pizza weight and will account for about one-third the calories provided by the pizza. To compare fat content, just divide fat weight given by ounces in a serving to get fat weight per ounce.

EXAMPLE:

ounces/serving = 4 fat weight = 11 grams
fat/ounce = 11/4 = 2.75 grams

ounces/serving = 5.5 fat weight = 19 grams
fat/ounce = 19/5.5 = 3.45 grams

When the serving weights of brands are the same, no calculation is necessary for the comparison.

All the pizza brands I have looked at represent fine nutritional value. If you are not very much concerned about additives, I suggest you choose your brand based purely on its taste, although some of the pizza manufacturers ought to be asked by consumer letters to remove the additives that they don't need to use.

If you are concerned about your weight, there are some pizzas that are lower in calories than others. Compare one product sold to those who are watching their weight, to a higher calorie product:

A Pizza Sold for Weight Control		**Regular**	
Per Ounce		*Per Ounce*	
Calories	56	Calories	69
Protein (grams)	3.3	Protein (grams)	4.2
Fat (grams)	2.4	Fat (grams)	2.94
Carbohydrate (grams)	5	Carbohydrate (grams)	7.6

Ounce for ounce, the weight control pizza is lower in calories, fat, protein, and carbohydrates than the other. The calorie savings is 13 per ounce — at least 50 calories per serving or as much as 92 per serving, depending on how much you eat.

□ SUBSTANCES ADDED TO PIZZA □

MATERIAL ADDED	LEGAL STATUS *(and GRAS class)*	PURPOSE
Dextrin	GRAS (1)	Texture control of sauce
Modified food starch	Additive (1–5)*	Texture control of sauce
Mono- and diglycerides	GRAS (1)	Emulsifiers
Oleoresin paprika	GRAS (food extract)	Coloring
Potassium bromate	Additive	Dough improver
Potassium sorbate	GRAS (1)	Preservative
Sodium aluminum phosphate	GRAS (1)	Baking acid
Sodium citrate	GRAS (1)	Buffer
Sodium metabisulfite	GRAS (1)	Preservative
Sodium stearoyl-2-lactylate	Additive	Dough conditioner
Xanthan gum	Additive	Sauce thickener

*Modified food starches are not now specifically identified on the label. They may be anything from class 1 to 5, depending on which starch is used.

□ FROZEN ENTREES □

There are two notable deficiencies in the labeling of frozen entrees: (1) You have no idea of the nutritional value or composition since most do not have nutritional labels. (2) You do not know how much you get of each ingredient.

While it is perfectly true that most people do not calculate the nutritional value of the dinner they may make at home, they can if they wish, since they know how much of each kind of food they use.

There is one overwhelmingly positive advantage of frozen entrees: they are convenient. Some even taste pretty good.

If you read the ingredients statement, some of the products do show ham or chicken or turkey as the first ingredient, meaning there is more of that than anything else, but it doesn't tell you how much. This is more reassuring than a label that lists water first (this is a cheap shot—the product listing water first included gravy, but it does illustrate that we just can't tell what we are getting).

Sifting through a number of labels, the following ingredients turn up in some:

Modified cornstarch: remember, that's starch that may have been treated in one of many ways. It's an additive, not GRAS.

Monosodium glutamate: the flavor enhancer—GRAS.

Hydrolyzed plant proteins: made from soybean and/or wheat protein that is broken down into amino acids by acids or enzymes and dried. It has a meatlike flavor and is used to increase meatlike taste and/or to save meat. It is GRAS.

Caramel coloring: used to make the product look as though it was made with more meat than it is. It is GRAS.

Autolyzed yeast: that is, yeast in which the cells have been burst by heating. It is used as a seasoning to increase meatlike taste. It is GRAS.

Turmeric: used as a color; GRAS.

Modified food starch: could be corn or wheat or tapioca or any other, but is an additive.

Oleoresin paprika: used as a coloring agent; GRAS.

Algin: a seaweed extract used to adjust the consistency of the sauce and prevent its curdling under freezing and thawing; GRAS.

Soy-protein concentrate: a food, extracted from soybeans, added for more meatlike texture at lower cost than meat. High nutritional value.

Comminuted turkey: beats me. Comminuted means chopped up fine. I am a little puzzled about why this would be done; maybe the shreds of the carcass are comminuted to add low-cost turkey taste.

Oleoresin turmeric: for color. Oleoresin is simply a concentrated extract. It is GRAS.

One tuna-noodle label lists:

"Milk, cooked noodles, tuna, mushrooms, celery, margarine (partially hydrogenated soybean oil, skim milk, salt, lecithin, mono- and diglycerides, sodium benzoate and citric acid [preservatives], artificial flavor and color, vitamin A palmitate and beta-carotene), bleached wheat flour, bread crumbs (unbleached enriched wheat flour [malted barley flour, niacin, iron, thiamine mononitrate, riboflavin], water, corn syrup, partially hydrogenated vegetable shortening [soybean oil], salt, yeast, nonfat dry milk, mono- and diglycerides, butter, honey, calcium propionate), salt, monosodium glutamate, artificial color, pepper."

This label is particularly clear, showing the ingredients of ingredients on the label. This label is informative because you can at least tell that there is more milk than cooked noodles, and more cooked noodles than tuna, and so forth. It is also clear that the sodium benzoate is used to preserve the margarine, not the frozen product, and that the calcium propionate (which we know is used to preserve bread) is used to make the bread out of which the bread crumbs are made.

The same company's chicken-and-noodle package lists the following ingredients:

"Milk, cooked noodles, chicken, mushrooms, chicken fat, enriched wheat flour, celery, salt, bread crumbs, monosodium glutamate, margarine, sugar, dehydrated onions, pepper, turmeric, natural flavorings."

For some reason, this label does not list the ingredients in the bread crumbs and margarine!

□ FROZEN VEGETABLES □

There are two kinds of frozen vegetables: those simply preserved by freezing with nothing added; and those prepared with some type of sauce or in a combination.

At present there are standards for only one vegetable: frozen peas. The standards describe quality, size, and physical defects primarily.

The most important things that decide the quality of the simple frozen vegetables are how they are frozen, stored, transported, and sold, and stored in the home. The more quickly a vegetable is frozen, and the lower the temperature is, the more closely it will resemble the fresh product, if it is not mistreated during storage and shipment. The best way to tell if the vegetable you buy has been subjected to rise and fall of temperature is to see how much ice there is rattling around loose inside the package. Under perfect conditions (rarely, if ever, achieved), there should be no free ice. And the more there is, the worse the situation has been. Some producers exert better control over shipping and storage conditions than others do. This is best reflected in the quality of the product's flavor and texture and the amount of free ice in the package.

It is technically possible to have a dye in packages that turns color to indicate that the product has thawed, but this innovation would be dynamite: it could ruin the frozen-foods industry without doing a great deal of good. (How could we tell if only the package itself rose in temperature, but the product has never thawed?)

Control over shipping and storage has improved over the past ten years, as have freezers in the stores.

A problem with the frozen-vegetable business is that there isn't much profit in it for the food processor. The business is highly competitive, and there is very little opportunity for product differentiation — why should you buy Brand X when it is not different from Brand Y, except in price?

To improve that situation from the processor's point of view, all kinds of prepared frozen vegetables have been introduced. Here are listings from three labels that illustrate those products:

Brand A
Peas in Butter Sauce

INGREDIENTS: Baby early June peas, frozen in a sauce containing water, AA butter, sugar, salt, cornstarch modified, and artificially colored with carotene.

Nutrition Information per Serving

Serving size	1 cup
Servings per container	1½ (or 4 2½-ounce servings)
Calories	150
Protein	7 grams
Carbohydrate	20 grams
Fat	5 grams

Brand B
Brussels Sprouts in Cheese Sauce

INGREDIENTS: Brussels sprouts in a sauce containing water, modified whey solids, partially hydrogenated soybean and cottonseed oils, cheese solids (cheddar and Parmesan), cornstarch modified, nonfat dry milk, flour, salt, pasteurized processed cheese, di- and trisodium phosphates, sodium alginate, sodium hexametaphosphate, mono- and diglycerides, sodium citrate, buttermilk solids, hydrolyzed vegetable protein, monosodium glutamate, onion powder, natural flavors, artificial colors (including F, D and C no. 5 [yellow]), garlic powder, lecithin, artificial flavors, disodium inosinate, and disodium guanylate.

Nutrition Information per Serving

Serving size	1 cup
Servings per container	1¼
Calories	170
Protein	6 grams
Carbohydrate	19 grams
Fat	7 grams

Brand C
Onions in Cream Sauce

INGREDIENTS: Whole onions, frozen in a sauce containing water, pasteurized process cheese, partially hydrogenated soybean oil, cornstarch modified, nonfat dry milk, salt, sugar, natural flavor, hydrolyzed vegetable protein, monosodium glutamate, onion powder, lecithin, and artificial color.

Nutrition Information Per Serving

Serving size	1 cup
Servings per container	1¼ (or 4 2½-ounce servings)
Calories	140
Protein	3 grams
Carbohydrate	14 grams
Fat	8 grams

The price of the peas with butter sauce was 30 percent higher than a package of plain frozen peas, although a comparison of brussels sprouts products showed the price to be only 15 percent higher for the product with butter sauce.

Our old friends modified cornstarch and sodium phosphates serve, as usual, to regulate the consistency of the sauce, and in Brand B to prevent water from migrating from sprout to sauce.

The mono- and diglycerides are there to make the sauce smoother, without using more fat.

The hydrolyzed vegetable protein, MSG, disodium inosinate, and disodium guanylate are strictly flavor contributions.

The net nutritional result is either good or bad, depending on how you look at it. Brussels sprouts with cheese sauce that are eaten, even though they contain fat, are more valuable than pure brussels sprouts never consumed at all. If the only way you can get your child to eat green vegetables is to buy stuff with sauce in it, the results may well be worth it. But you do double the calories by adding the fat.

It is distressing to see artificial colors used, although it might cost a bit more to get an attractive color without the use of F, D and C number 5 (yellow).

One excellent advantage of the frozen packaged vegetables

— with or without sauces — is that you can use the nutritional information on the label to plan and help your diet.

You should be consuming at least one half cup each of two kinds of vegetables every day as well as one potato. If those vegetables were spinach and lima beans, a look at the nutritional label would show you that these two portions will supply 150 percent of the vitamin A and 75 percent of the vitamin C you need each day, plus many other nutrients.

If you prefer fresh vegetables — and many of us do — read the labels of the frozen products to learn what you will be getting nutritionally in the fresh vegetable; there is no loss of nutrients worth calculating. In other words, use the frozen-vegetable package as your "reference library" for the fresh product. (Indeed, the likelihood is that the frozen product has *less* vitamin loss than the fresh since it is subjected to higher temperatures for less time after harvest.)

□ SUBSTANCES ADDED TO FROZEN ENTREES, SPECIALTIES, AND VEGETABLES □

MATERIAL ADDED	LEGAL STATUS *(and GRAS class)*	PURPOSE
Algin	GRAS (2)	Thickener
Autolyzed yeast	GRAS (1)	Seasoning
Caramel color	GRAS (1)	Coloring
Citric acid	GRAS (1)	Preservation aid
Di- and trisodium phosphate	GRAS (1)	Texture modifier
Disodium guanylate	Additive	Flavor enhancer
Disodium inosinate	Additive	Flavor enhancer
F, D and C #5 (yellow)	Color Additive	Coloring
Hydrolyzed plant protein	GRAS (1)	Seasoning
Lecithin	GRAS (1)	Emulsifier
Modified cornstarch	Additive (1–5)*	Thickener
Modified food starch	Additive (1–5)*	Thickener
Mono- and diglycerides	GRAS (1)	Emulsifiers

*Modified food starches are not now specifically identified on the label. They may be in any class from 1 to 5, depending on which starch is used.

MATERIAL ADDED	LEGAL STATUS *(and GRAS class)*	PURPOSE
Monosodium glutamate (MSG)	GRAS (1)	Flavor enhancer
Oleoresin paprika	GRAS (food extract)	Coloring
Sodium alginate	GRAS (2)	Thickener
Sodium benzoate	GRAS (1)	Preservative
Sodium citrate	GRAS (1)	Acidity control
Sodium hexameta-phosphate	GRAS (1)	Texture modifier
Soy-protein concentrate	GRAS (1)	Seasoning
Turmeric	GRAS (spice)	Coloring

Chapter 12

CANNED GOODS

Vegetables, Fruits, Soups

Since the introduction of frozen foods, canned goods sales have been declining. Still, most of us eat canned fruits, vegetables, and soups.

Because canning is in itself a preservation method, it is not common to find any preservatives in canned goods. Indeed, in the basic canned fruits and vegetables, one rarely finds anything added except some form of sugar or salt. However, as in many other parts of the food industry, manufacturers do try to improve their profits by producing specialties that add value, in the form of convenience, flavor, or nutrition, to the basic food commodity, and for these products ingredients become more complex.

Fruits and vegetables when canned are by and large standardized foods, although canned soups are not.

□ VEGETABLES □

Under the standards, the ingredients listed in the following chart may be used and must be declared.

☐ SUBSTANCES ADDED TO CANNED VEGETABLES ☐

Material Added	Legal Status (and GRAS class)	Purpose
Ascorbic acid	GRAS (1)	Preservative (antioxidant)
Autolyzed yeast extract	GRAS (1)	Flavoring
Baking soda	GRAS (1)	Acidity adjustment
Calcium salts	GRAS (not classified)	Firming agent
Citric acid and other acids	GRAS (1)	Acidity adjustment
Disodium guanylate	Additive	Flavoring
Disodium inosinate	Additive	Flavoring
EDTA	Additive	Preservative (sequestrant)
Flavoring (except artificial)	GRAS (not classified)	
Hydrolyzed vegetable protein	GRAS (1)	Flavoring
Monosodium glutamate (MSG)	GRAS (2)	Flavor enhancer
Salt	GRAS (4)	Flavoring
Spice	GRAS (not classified)	Flavoring
Sugar	GRAS (2)	Flavoring

Not all of the permitted ingredients are permitted in all vegetables. For example, calcium salts are permitted in canned tomatoes but not in string beans. Calcium salts may be necessary to prevent the tomato from breaking apart but not to keep the physical integrity of a string bean.

Almost all canned vegetable brands are nutritionally labeled.

While there is no weight standard for the amount of vegetable and the amount of water in the can, the test for amount of fill indirectly controls the proportions. For example, if a can of peas is emptied and refilled, the peas must come up to the top of the can and stay there after 5 minutes. That is, the liquid does not float the peas, but takes up the empty space between them.

The standards for the quality of the vegetables are also very specific, covering thirty-three printed pages, and setting limits for whole and broken pieces or kernels, absence of stems, and so forth.

The label must indicate whether a vegetable is seasoned with spice or other ingredients, as in the name "sweet corn seasoned with butter," for example. If the quality standard is not met, for instance, if canned corn contains more corn silk than the standards allow, then the label must say "excessive silk."

The use of acids is very important not only for flavor but for assuring safety and stability of the canned vegetable. Vegetables have a natural and *varying* acidity. Vegetables such as corn or peas or string beans are often too alkaline, not acid enough. This means that the vegetable would require longer heat treatment, with consequent decline in texture, to kill all the bacteria. Acid may be added to increase the safety margin and reduce the sterilizing time.

EDTA is permitted in some vegetables, as is ascorbic acid. EDTA prevents traces of metal from discoloring the product and spoiling the taste, and ascorbic acid maintains a fresh, bright color.

The true differences in quality of canned vegetables are not particularly evident from reading the label. So much depends on the strain of vegetable used, the harvesting and transportation practices, the processing methods, and the degree of quality control in the cannery. For example, one tomato may not require calcium salts to remain firm during the canning; another variety might. The presence of calcium salts on the label doesn't really tell you whether one tomato is firmer than the other. The only way to tell is to try them. There is nothing on the label to make you want to choose one brand rather than another because of taste or texture. The presence or absence of added materials does not reveal superior or inferior taste or texture or color.

□ FRUIT □

There are standards for many of the canned fruits, whether they are packed in syrup or their juice, or artificially sweetened. The standards control the quality of the fruit, the ingredients that may be added and that must be declared, and all aspects of labeling.

The words "extra light," "light," "heavy," and "extra heavy syrup" are to be used when a certain amount of sugar is used in the syrup. For example, for peaches, the percentage of sugar allowed is:

Extra light: 10–14
Light: 14–18
Heavy: 18–22
Extra heavy: 22–35

About 40 percent of the weight in a can of heavy-syrup-packed fruit is syrup. Calories can run from under 100 per cup to over 200 per cup. Almost all cans of fruit are nutritionally labeled. Canned fruit can be an important source of vitamins, fiber, and minerals.

The amount of fruit used need not be on the label, but is stated by some processors voluntarily. For example, a label for one brand of yellow cling peaches reads: "Net weight 8¾ ounces, weight of peaches 5½ ounces before addition of water for processing." This sort of statement should be legally required, but it is not, and thus people may be misled by a lower-priced product, which actually contains more sugar and water and less fruit. Look for brands that do state the amount of fruit.

The optional ingredients permitted in canned applesauce are: water; salt (GRAS: 4); organic acids (GRAS: 1); sugars (GRAS: 2); spices (GRAS); natural and artificial flavoring (GRAS and additives); erythorbic acid (GRAS: 1, as an antioxidant); ascorbic acid (GRAS: 1, as a nutrient); color additives (permitted "as long as they don't make the product look better than it is." Why add them at all?); and pectin (GRAS: 1), in artificially sweetened products.

Most other standardized canned fruits usually do not permit all these materials; only natural and artificial flavors, vinegar, lemon juice or other organic acids, and spices are allowed and must be clearly labeled.

The quality of canned fruit is also affected more by the quality of the fruit and the processing conditions than by any of the added ingredients. The only product about which there might be some concern is maraschino cherries, sometimes found in canned fruit salad. The element of concern here is the dye. Two previously approved dyes have since been prohibited. It might interest you to know that the only thing a maraschino cherry has in common with a real cherry is the shape. They are made by packing cherries in sulfur-dioxide brine, bleaching all the color out, removing all the natural juices and sugars, and ending up with a pale-yellow, round, cherry-shaped object, with a nice bite to it, which is then impregnated with red dye, sugar, and benzaldehyde, an artificial flavor. That doesn't mean they don't taste good, but they are hardly one of nature's pure, beautiful fruits. Think of them like candy—only more artificial.

□ CANNED SOUP □

There are no standards for canned soups, but something that I consider far better: a long history of outstanding performances by a few food processors in producing wholesome, nutritious, good-tasting products, made as simply as it is possible to make them and still have them taste good and remain stable. Historically, the major soup packers have spent millions to research crops, vegetable breeds, and processing methods to obtain high-quality soup.

I personally have been eating one brand of tomato soup since 1925 and, incredibly, it has not varied in taste to me, although I am sure there must have been some changes in composition.

As a food source, soups have a very unusual function: they are highly satisfying, low in calories, low in fat calories, and, depending on the type, good sources of nutrition, even of protein in the case of the meat and chicken soups. Soups, whether canned or homemade, can play a key role in a low-calorie diet because of these characteristics. Here again, the label can supply the information we need to make a choice.

Compare the ingredient listings for two brands of beef barley soup.

Brand A

INGREDIENTS: Water, beef, barley, carrots, tomatoes, concentrated beef stock, food starch — modified, salt, dehydrated potatoes, celery, hydrolyzed vegetable protein (with caramel, yeast extract, hydrolyzed peanut oil, disodium inosinate, disodium guanylate), peas, beef fat, cornstarch, corn syrup, caramel coloring, monosodium glutamate, onion powder, natural flavoring, garlic power.

Brand B

INGREDIENTS: Beef stock, beef, tomatoes, barley, carrots, potatoes, celery, water, salt, cornstarch, peas, potato starch, hydrolyzed plant protein, wheat flour, caramel color, monosodium glutamate, and natural flavoring.

Nutrition Information per Serving

Serving size	5½ ounces—condensed (11 ounces as prepared—312 grams)
Servings per container	2
Calories	110
Protein (grams)	7
Carbohydrate (grams)	15
Fat (grams)	3

Percentage of U.S. Recommended Daily Allowances (RDA)

Protein	10	Riboflavin	2
Vitamin A	15	Niacin	6
Vitamin C	2	Calcium	*
Thiamine	2	Iron	4

*Contains less than 2 percent of the U.S. RDA of this nutrient.

The label for Brand A is a testimony to carelessness in the use of ingredients as well as in the label preparation. (There is no such thing as "hydrolyzed" peanut oil.) Brand A contains two additives not used by B, disodium inositol and guanylate, and in addition, where B uses the food "cornstarch," A uses the additive "food starch — modified" as the principal texture regulator, plus cornstarch.

It is not unusual to find cheaper products using more additives. Beef and beef broth and chicken are expensive. Often MSG and other flavor enhancers are used to make less taste like more.

Finally, Brand A does not have a nutritional label and B does. The price difference between the two was three cents.

Now compare labels on two split-pea soups.

Brand C

INGREDIENTS: Water, split green peas, smoked ham (cured with water, salt, sugar, sodium tri-, meta-, and pyrophosphates, sodium nitrite, flavoring), carrots, onions, celery, salt, sugar, hydrolyzed plant protein, monosodium glutamate, autolyzed yeast, spices, dextrose, partially hydrogenated vegetable oil, natural smoke flavors, disodium inosinate, disodium guanylate, and natural flavors.

Brand D

INGREDIENTS: Water, green split peas, ham, carrots, salt, rendered pork fat, sugar, dehydrated onion, flavorings, hydrolyzed vegetable protein.

This is interesting. Brand C appears to have far more substances added, but this may be an illusion. Brand D simply declares "ham," while C declares all the ingredients in the ham as well. D's ham probably contains the very same ingredients — indeed, it must, to be ham — but they are simply not declared. This means that Brand D's label does not meet the legal requirements.

In addition, however, C does contain a number of ingredients that D doesn't — hydrogenated vegetable oil, natural smoke flavors, disodium inosinate, disodium guanylate.

If I liked both products equally, I would select D as the brand to continue to buy.

Let me illustrate the value of the nutritional label. This list appears on a can of cream of chicken soup.

Nutrition Information per Serving	
Servings per container	2
Calories	140
Protein (grams)	4
Carbohydrate (grams)	10
Fat (grams)	9

With 9 grams of fat and a total of 140 calories per serving, 81 calories — 58 percent of them — come from the fat. (Nine grams times 9 calories per gram equals 81; 81 divided by 140 times 100 equals 58 percent.) This is higher than the recommended 35 percent fat calories, which means that if I were working to lower my fat-calorie contribution, or if I were on a reducing diet, I would not use that particular soup variety. The beef barley soup would be a better choice. With only 3 grams of fat, 27 fat calories are contributed, only 24 percent of the 115 calories in a serving of that soup.

Those little hand computers will continue to be very helpful until food processors state the calorie percentage from fat — as the smarter and more reliable processors are already doing!

■ A WORD ABOUT DRY SOUP MIXES

Dry soup-mix ingredients are similar to those in canned soups. A few, however, are less common in canned soup. They are: mono- and diglycerides, used to help disperse the ingredients when water is added; antioxidants like BHA, citric acid (GRAS), and propyl gallate (GRAS); and thickeners such as modified food starches (additives) or gums (GRAS).

Nutritional labels are not found on dry soup mixes in most cases. These mixes basically make a pleasant salty beverage, unless they are made with noodles. The noodles in dry soup mixes are superior in texture to those in canned soups. Noodles don't go through the canning process retaining much integrity of texture.

The full range of flavor enhancers and meat-flavor sources such as MSG, disodium inosinate, disodium guanylate, hydrolyzed vegetable protein will be found in most dry soups. The vegetables are of course dehydrated. Many people use dry soups as seasoning mixes to flavor other soups or foods.

□ SUBSTANCES ADDED TO SOUPS □

Material Added	Legal Status *(and GRAS class)*	Purpose
Autolyzed yeast	GRAS (1)	Flavoring
BHA	GRAS (3)*	Antioxidant
Caramel coloring	GRAS (1)	Coloring
Citric acid	GRAS (1)	Acidifier, antioxidant
Dextrose	GRAS (2)	Sweetner
Disodium guanylate	Additive	Flavor enhancer
Disodium inosinate	Additive	Flavor enhancer
Food starch modified	Additive (1–5)†	Thickener
Hydrolyzed plant or vegetable protein	GRAS (not classified)	Flavoring
Mono- and diglycerides	GRAS (1)	Emulsifiers
Monosodium glutamate (MSG)	GRAS (2)	Flavor enhancer
Natural smoke flavors	Additive	Flavoring
Propyl gallate	GRAS (1)	Antioxidant
Salt	GRAS (4)	Flavoring
Sodium metaphosphate	GRAS (1)	Holds water in ham
Sodium nitrate	Additive	Preservative
Sodium pyrophosphate	GRAS (1)	Holds water in ham
Sodium triphosphate	GRAS (1)	Holds water in ham
Yeast extract	mislabeled	?

*GRAS (3) in fats and oils. Otherwise an additive.

†Modified food starches are now specifically identified on the label. They may be anything from class 1 to 5, depending on which starch is used.

Chapter 13

SALAD DRESSINGS, OILS, PEANUT BUTTER, JAMS, AND SAVORY SNACKS

□ SALAD DRESSINGS □

The term "food dressings" is used by the FDA to cover French dressing, mayonnaise, and salad dressings. As a food group, these products serve the very useful purpose of making fresh vegetables more appetizing — at a cost of high calories, high fat, and very little nutritional value in themselves.

I am not suggesting that we eliminate such dressings, but rather, we should look for the products that provide the most flavor benefit at the lowest calorie and fat cost. There are many of them. Reading and understanding food dressing labels is well worthwhile if we are to reduce our fat consumption.

While it is true that vegetable oils are a source of vitamin E and, in some special cases, unsaturated fats, high-fat dressings are not really necessary as a source of these nutrients; we get them or can get them in other, more nutritionally rich foods. Most of us are getting all the vitamin E we need and more from the excessive amount of fat we eat.

■ STANDARDS FOR FOOD DRESSINGS

At the time standards were established, the main purpose was to make sure that the consumer got at least the amount of fat and eggs they thought they were getting. Mayonnaise contains at least 65 percent fat and salad dressing about 30 percent, and the FDA as well as consumer groups wanted to protect the consumer from buying something that might be called "mayonnaise" that did not contain the required amount of fat and eggs.

But times have changed, and now we know that less fat and fewer eggs are better than more — at least as far as cholesterol control is concerned. (More than 90 percent of the calories in mayonnnaise comes from fat.) Therefore, it is no longer considered unfavorable to the consumer to call a low-fat product "mayonnaise." Hence the term "low-calorie mayonnaise" is permitted today. (When making low-calorie mayonnaise, the vitamin-E content is boosted to the amount that would be found in standard mayonnaise.)

Take a look at the comparative analyses (page 202) taken from the labels of a variety of food dressings. All figures are for one tablespoonful — the standard serving size.

In opacity and apparent viscosity, the imitation mayonnaise, mayonnaise, Mayonette, and Miracle Whip look alike, which is made possible by certain additives. Their colors are slightly different.

■ ADDITIVES USED IN FOOD DRESSINGS

Normally, mayonnaise can be made by blending oil, vinegar, spices, egg yolks, a bit of salt. To make low-calorie mayonnaise, we must remove much of the oil. When this is done, there is too much vinegar acidity, so we must add some water to dilute the acid. But when we add the water, the acidity is not high enough to delay mold growth, so we must add a preservative — a sorbate or a benzoate. Finally, to protect against off flavors owing to contamination by traces of metal, some manufacturers add a sequestrant, which combines with

□ COMPARATIVE ANALYSES OF FOOD DRESSINGS □

Per One Tablespoon

PRODUCT	CALORIES	FAT *(grams)*	CARBOHYDRATE *(grams)*	FATS *Polyunsaturated (grams)*	FATS *Saturated (grams)*	*Cholesterol (milligrams)*
Imitation Mayonnaise	40	4	1	3	1	5
Miracle Whip	70	7	2	4	1	5
Mayonette	33	3	2	no statement		
Creamy Cucumber	30	3	1	no statement		
Mayonnaise	100	11	0	6	2	5

the metal and prevents it from reacting with the other ingredients. EDTA is a sequestrant.

You can avoid these additives by continuing to eat real mayonnaise but using less of it. If you can make half a tablespoonful do what a full one did, there go half the calories (and cost).

Real mayonnaise brands do not vary much in their ingredients, except that some contain EDTA and others do not. This proves that EDTA is not necessary to produce mayonnaise.

However, when it comes to imitation mayonnaise and liquid salad dressings, quite a few nonfood materials are found. Look at these ingredient statements:

Brand A
An Imitation Mayonnaise

INGREDIENTS: Water, soybean oil, food starch modified, egg yolk, vinegar, salt, sugar, mustard flour, citric acid, sodium benzoate, potassium sorbate and calcium sodium EDTA, natural flavor and color, BHA and BHT.

Brand B
A Liquid Salad Dressing

INGREDIENTS: Water, corn syrup, vinegar, soybean oil, egg yolk, food starch modified, cellulose gum, salt, polysorbate 60, potassium sorbate and calcium disodium EDTA (as preservatives), natural and artificial flavorings.

Some new ingredients turn up in one brand of liquid dressing: xanthan gum and propylene glycol alginate. These materials are used to achieve the texture and flavor of high-fat salad dressings, using much less fat. Xanthan gum, as we have seen elsewhere, is a thickener used to regulate consistency. This is also true of propylene glycol alginate. Manufacturers try to obtain the desired thickness without pastiness and without hiding flavor. To achieve this, they must use a wide range of materials that modify consistency, rather than only one.

Another salad dressing contains monosodium glutamate and oxystearin in addition to the food ingredients. Oxystearin prevents the fat from crystallizing in the refrigerator. It is an additive, not GRAS. Monosodium glutamate (GRAS), is added for flavor.

☐ SUBSTANCES ADDED TO FOOD DRESSINGS ☐

MATERIAL ADDED	LEGAL STATUS *(and GRAS class)*	PURPOSE
Acetic acid	GRAS (1)	Acidifier (vinegar substitute)
BHA	GRAS (3)*	Antioxidant
BHT	GRAS (3)*	Antioxidant
Calcium sodium EDTA (and EDTA)	Additive	Preservative (sequestrant)
Citric acid	GRAS (1)	Acidifier
Modified food starch	Additive (1–5)†	Thickener
Monosodium glutamate (MSG)	GRAS (2)	Flavor enhancer
Oxystearin	Additive (1)	Prevents oil from crystallizing in refrigerator
Polysorbate 60	Additive	Emulsifier (to keep oil droplets small)
Potassium sorbate	GRAS (1)	Preservative
Propylene glycol alginate	Additive (2)	Thickener
Sodium benzoate	GRAS (1)	Preservative
Xanthan gum	Additive	Thickener

*GRAS (3) in fats and oils. Otherwise an additive.
†Modified food starches are not now specifically identified on the label. They may be anything from class 1 to 5, depending on which starch is used.

Many dressing products do not have nutritional information on the label. This is for one of two reasons. Either there is not enough room, or the news is not good! My guess is that when a food can show some "good" nutritional information, a way is found to label it.

What about the polyunsaturated-fat claims? In all cases where stated, the ratio of polyunsaturated fats to saturated is better than the American Heart Association recommendation, which you may remember from chapter 5 is 1.4 to 1. We simply don't know the ratio where there is not a statement on the label unless we look up the figures for the oil itself.

The following chart shows the *percentage* of saturated and unsaturated fat for each of the common oils and the ratio of unsaturated to saturated fats.

	POLYUNSATURATED	SATURATED	RATIO
Corn	53	10	5:1
Cottonseed	50	25	2:1
Olive	7	11	1:1.6
Peanut	29	18	1.6:1
Safflower	72	8	9:1
Sesame	42	14	3:1
Soybean	52	15	3.5:1
Sunflower	64	21	3:1

From A. L. Merrill, B. K. Watt, et al, *Composition of Foods* (Agricultural Handbook no. 8).

None of the above figures applies to hydrogenated oils. These figures are only for the refined, *un*hydrogenated oils.

□ MORE ABOUT OIL □

Not very many years ago, the vegetable shortening that we bought was a white, plastic material. It was hydrogenated and very high in saturated fats. With the growing knowledge of nutrition and concern about saturated fats, cholesterol, and heart disease, the food industry has modified its products so that, even in some baked foods, oils are often used rather than the hard fats.

Hydrogenation is still performed, but only enough to stabilize the fat to oxidation, that is, to protect against rancidity. The common term used today is "partially hydrogenated."

For home-baking purposes, such oils frequently have "polyglycerides added," which are emulsifiers useful in cake baking.

Frying fats used commercially are still highly hydrogenated and very low in unsaturated fats. This is necessary to protect the fat against fast breakdown in frying and to prevent greasiness in the fried food.

It is important to know that many processed foods — fried and frozen — continue to use hydrogenated fat. These fats typically contain:

	% POLYUNSATURATED	% SATURATED
Vegetable Frying Fat	7	23
Animal Frying Fat	11	43

■ VINEGARS

Vinegar is principally acetic acid and water produced by bacteria fermenting any one of a wide variety of carbohydrate- or sugar-containing materials. The raw materials might be wine, cider, malted wheat, malted barley, rice, or just plain sugar.

The price of vinegar ranges from 51¢ a quart or less for distilled white, private-label vinegar, up to $20.00 per quart or more. Even in many supermarkets, you can spend $5.00 per quart.

Vinegar has no direct food value of any kind other than a few calories (about 125 per pint, so few they can be ignored), contains no vitamins, no minerals, and is nutritionally useless, although the acetic acid is handled as carbohydrate by the body. Most vinegars are standardized to contain 5 percent acetic acid.

The discovery of vinegar is one of the more important food discoveries made by man. There are references to vinegar among Roman authors dating back to 200 B.C. It was called "acetum," from which acetic acid gets its name. Most disease-producing bacteria cannot live in an acid environment, and thus vinegar was used not only to prevent spoilage and loss of food, but undoubtedly prevented many food-borne illnesses as well.

But most important is the effect of vinegar on taste — and that is why we see such a wide price range. The price varies with the vinegar source. Any sugar can be converted to alcohol by yeast and then fermented to vinegar. Grain alcohol can be used as a source to produce cheap vinegar; or fine wine can be used, with or without the infusion of fresh herbs, to produce a very expensive product. If you use a flavorful wine vinegar, you should be able to cut way down on the amount of oil you use in salads and still have tasty dressings.

□ PEANUT BUTTER □

Peanut butter is essentially peanuts that have been roasted and ground. This material tends to separate as the oil rises to the top, and is rather difficult to spread at temperatures lower or higher than room temperature.

Peanut butter is an important food. We consume more than 500 million pounds per year — that is more than 2 pounds for every man, woman, and child. Consumption jumped when peanut butter was first stabilized to prevent oil separation, around 1940.

Peanuts can be affected by the poison aflotoxin, which comes from mold. Unusually careful steps are taken to be sure that peanuts are free of this mold, and all peanuts with more than 30 parts per *billion* are rejected for food use.

Peanut butter, with no additives, is 27 percent protein and 49 percent fat. Peanut butter with sugar, hydrogenated vegetable oil, and salt added contains 25 percent protein and 53 percent fat, which shows that about 4 percent of fat is added. Another brand shows only 50 percent fat and 26 percent protein. From this information, one can conclude that the material added to the simple butter may run between 2 and 5 percent fat. None of this is significant nutritionally: the amount added is too small to matter (maximum additions total 1 to 2 grams in a full ounce).

No brand of peanut butter examined showed the use of any preservatives, but some stated "no preservatives." Peanut butter, therefore, should be chosen purely on the basis of price, taste, and texture preference.

I noticed that the "processed" peanut butters all had nutritional labeling, but those claiming to be all natural did not! This makes nutritional comparison impossible in the stores, but research figures show no significant nutritional difference.

It is important to realize that peanut butter provides 70 percent of its calories from fat — far higher than the recommended 35 percent level — but fortunately, man doesn't live by peanut butter alone. Bread helps reduce the percentage of calories from fat, and even the jelly we eat in omnipresent peanut-butter-and-jelly sandwiches helps.

□ JAMS AND JELLIES □

These foods are standardized. They must be at least 45 percent fruit, and not more than 55 percent sugar. In the case of fruit butters, the ratio is different. Fruit butters contain a lot more water than jams and jellies: about 50 percent water, as compared with 30 percent in jams and jellies. Originally the standards were set to make sure that nothing was called jam or jelly unless it contained *at least* a certain amount of fruit and that the ingredients were limited to only those approved for use in the standardized food. However, while this is still true, there are now a growing number of fruit spreads with less sugar, to help people meet their desire for continued enjoyment with less sugar.

When reading jam, jelly, and fruit-spread labels, the most important thing to look for is the calorie statement, if present. If the product is called jam or jelly or preserve and does not have a nutritional label on it, you can safely assume that it contains 18 calories per teaspoonful, no matter what flavor. If it is a fruit spread called by some other name and claims lower calories, it will have to state the number of calories.

Jams and jellies are simply not an important source of nutrition alone, but they are important to our enjoyment as well as to complement peanut butter and a host of other foods. Do not expect any contribution to vitamins, protein, or minerals, although there may be some fiber contribution in the case of jam or preserves.

In general, the price of a jam is related to the number of ingredients on the label! The most expensive jams (and best-tasting, usually) have only two ingredients: fruit and sugar.

The less expensive products may contain a variety of ingredients besides fruit and sugar. They are:

Pectin: used to make the product gel. Pectin is extracted from apples or lemons and is GRAS.

High-fructose corn syrup: this is a relatively new ingredient, which is cheaper than cane or beet sugar and has a different type of sweetness. It can offer more sweetness for fewer calories; but do not assume this unless it is confirmed on the label.

Corn syrup: sometimes used to prevent a jam from crystallizing.

Citric acid: extracted from lemons or produced synthetically. It adjusts the acidity when the fruit isn't acid enough for taste.

Sodium citrate: a buffering agent, that is, it keeps the acidity constant and thus assures uniform acid taste.

Artificially sweetened jams and jellies may contain many other additives, especially preservatives to replace one function of the sugar: the more water and the less sugar there is present, the easier the product spoils by fermenting. One grape spread also contains potassium sorbate to retard spoilage after opening, to compensate for the reduced sugar content. Potassium sorbate is GRAS.

Like other high-sugar foods with little nutritional value, the use of jams and jellies should be limited to a moderate quantity. For those who are big jam and jelly eaters and who wish to reduce their sugar intake: these products should be on your list to eat less of or to replace with lower-sugar products.

□ SAVORY SNACKS □

A group of foods so many of us enjoy are the snacks that are not sweet. Usually, these products are served with dips and with beverages: cocktails, beer, or sodas.

Knowing what the additives are in these foods is not nearly as important as understanding their contributions to your diet, especially in relation to the dietary guidelines proposed to improve our nation's health. The guidelines are described in greater detail in the appendix, but three of them are very pertinent to this subject of salty or savory snack foods. They are:

1. Reduce your overall fat intake to contribute only about a third of your total calories.
2. Avoid obesity.
3. Reduce your sodium intake.

Well, the salty snacks, especially the way in which they are eaten,

increase your overall fat intake in most cases,
increase your caloric intake and promote obesity,
increase your sodium intake.

I include in this category of foods potato chips, fried corn-based products, extruded fried dry snacks, salted crackers, pretzels, bread sticks. All the fried products — potato chips, corn chips, cheese-flavored puffs — are 40 percent fat and contain from 160 to 170 calories per ounce.

Potato chips supply 170 calories in an ounce, and 63 percent of those calories comes from fat instead of the 33 percent most experts — especially the American Heart Association — recommend.

Unsalted bread sticks and crackers are much to be preferred, with under 110 calories per ounce.

Potatoes are an excellent food; however, potato chips are the least nutritionally useful way to eat them. As the nutritional comparison shows, you do get 10 percent of the vitamin C requirements from an ounce of potato chips, along with the

□ NUTRITIONAL COMPARISON □

	Potato Chips	Boiled Potatoes
Serving size (ounces)	1	4
Calories	170	85
Protein (grams)	2	2.3
Carbohydrate (grams)	13	18
Fat (grams)	12	0.1
Vitamin C (% RDA)	10	50
Niacin (% RDA)	6	7

170 calories, compared to 50 percent of the RDA of vitamin C in 85 calories' worth (4 ounces) of boiled potatoes.

The label of one brand of "natural" potato chips reads, "No Preservatives Used. Ingredients: Unpeeled potatoes, cottonseed or peanut oil, and salt." No nutritional labeling is supplied. The "natural" billing is irrelevant. This product increases your overall fat, calorie, and salt intake just like any other salted fried snack or chip. (Many brands of potato chips now do supply nutritional labeling.)

Some brands use hydrogenated oils, others do not. If you are going to eat chips, you will get more unsaturated fat from the unhydrogenated oils.

Pretzels are virtually fat-free but contain a little more salt. They certainly are lower in calories than chips: approximately 100–110 per ounce. Bread sticks are another low-fat choice for snacks. While they do not have nutritional labels (which is foolish, since nutritional information might increase their sales), bread sticks also contain 100–120 calories per ounce with very low fat — under 3 percent. Saltine-type crackers are as high as 12 percent fat and contain 125 calories per ounce, contributing 23 percent of the calories as fat — still very much lower than chips.

The added materials you might find in some of these snack products are disodium inosinate and disodium guanylate. Both are flavor enhancers similar to MSG (monosodium glutamate). Both are additives, not GRAS. (MSG is GRAS.)

Some brands of extruded fried snacks — cheese, onion, or sour-cream flavored — are nutritionally fortified to a small degree. I suspect that these foods will not meet the new fortification guidelines, although, clearly, if these foods are going to be eaten in quantity, it is probably better to eat fortified ones. However, in no case do I suggest using a fortified, 40-percent-fat product over a nonfortified, 3-percent-fat product as a snack.

But once again, these foods supply real pleasure. It is important to enjoy what we eat and to have foods that are primarily "fun" foods. How can we achieve this?

Well, one way is to use raw carrots, celery, bread sticks, rice crackers to carry dips, avoiding the high-fat chips while providing an enjoyable snack.

Another is to *earn the snack.* Let's say that you want to have some chips and dip, say 300 calories' worth. First you must realize that if you added 300 calories per day to your regular diet, you would gain one pound every 10 days — 3 pounds per month. But you can earn the snack by:

walking three miles at a good clip, or
jogging for half an hour, or
riding a bike for half an hour, or
swimming a quarter of a mile.

You can also *"buy"* the snack by foregoing other high-fat foods. For example, if you now drink two cups of milk per day, full fat, and you switch to skim milk, you will save 16 grams of fat — nearly enough to "buy" 1 ounce of chips.

Finally, if you now snack and are not overweight, don't pay very much attention to what I've said, except for two things: check the percentage of calories you are getting from fat and move to 33 percent; and cut down on sodium, that is, salt. The salt level is not now stated, so you are unable to use label information to control your salt intake. You can, however, avoid products that are coated or dusted with salt.

□ SUBSTANCES ADDED TO SAVORY SNACKS □

Material Added	Legal Status (*and GRAS class*)	Purpose
Artificial color	Color additive	Coloring
BHA	GRAS (3)*	Antioxidant
Citric acid	GRAS (1)	Acidifier
Disodium guanylate	Additive	Flavor enhancer
Disodium inosinate	Additive	Flavor enhancer
Disodium phosphate	GRAS (1)	Processing aid
Lactic acid	GRAS (1)	Acidifier
Monosodium glutamate (MSG)	GRAS (2)	Flavor enhancer
Propyl gallate	GRAS (1)	Antioxidant
Salt	GRAS (4)	Seasoning
Sodium bicarbonate	GRAS (1)	Alkalizer

*GRAS (3) when used in fats and oils. Otherwise an additive.

Chapter 14

THE THIRST QUENCHERS

There is a connection between carbonated soft drinks, fruit juices, fruit drinks, frozen concentrated juices, and powdered drink mixes. The connection is water, which is what ends up quenching our thirst.

□ FRUIT JUICES □

Let's start with the fruit juices. Compare the water, sugar, and calorie content of the common juices:

JUICE	% WATER	% SUGAR	CALORIES/CUP (8 OUNCES)
Apple	88	12	120
Orange	88	11	110
Grapefruit	81	9.6	95
Pineapple	86	13.5	135
Tomato	94	4.3	50
Grape	83	16.3	165
Apricot (nectar)*	85	14	140
Prune	80	19.5	200
Lemon	91	6	60
Lime	90	6.5	65

*Apricot puree sweetened with sugar syrup.
From J. G. Woodroof and B. S. Luh, *Commercial Fruit Processing* (Westport, Connecticut: AVI, 1975), pp. 161, 546.

Most of us don't drink fruit juices when we are very thirsty but rather as part of a meal — usually breakfast — and as a source of vitamin C. A four-ounce portion of orange juice provides the full recommended daily allowance of vitamin C, no matter whether it is freshly squeezed, refrigerated, or frozen.

■ CONCENTRATED JUICES AND PUREES

You will find the words "concentrated grape juice" or "orange concentrate" on labels of some products, or "grape juice from concentrate."

Large numbers of people and fruit don't necessarily grow in the same geographical area. Products containing lots of water are most economically produced and packaged near their largest markets so that the water doesn't have to be shipped for thousands of miles. Hence, fruit juice is often concentrated, that is, much of the water is removed either by evaporation or freeze concentration, and the concentrate is then shipped to the drink-packing plant where the water is replaced to give an approximation of the original juice.

Another advantage of this procedure is that the concentrate may be collected and stored at harvest time at enormous savings of space and money, which ultimately is reflected in the price of the product to the consumer.

Concentrates are either canned or frozen and are stored in that manner.

■ DRINKS AND PUNCHES

There are a variety of products that contain *some* fruit juice and that are called fruit drinks or punches or by a trade name. They must show the percentage of fruit juice on the label; they range from 8 percent to 50 percent fruit juice.

They are usually fortified with vitamin C so that the recommended serving contains 100 percent or more of the recommended daily allowance. Their sugar content and calorie value are about the same as those of the naturally sweet juices: about 15 calories per ounce and about 10 percent sugar.

If the fruit juices contained nothing more than water, sugar, and vitamin C, there would not be any difference between the "drinks" and the juices. But, of course, the juices contain many more known substances and possibly some unknown as well. What is *not* known is whether those substances are of nutritional or health value.

It is not unreasonable to expect the natural juices to be more expensive than their competing flavored drinks, but this is not always so! Prices vary all over the lot, and often the "drinks" are more expensive than the real thing.

For example, these were the prices in one supermarket on a given day (all in 32- to 46-ounce containers):

	Price per Ounce (cents)	% Juice
A Punch	1.4	15
Drink A	1.4	8
Pure Vegetable Juice	1.6	100
Pineapple Juice	1.4	100
Apple Juice	2.1	100
Tomato Juice	1.7	100
Drink B	2.7	50

■ DRINK MIXES

You can buy dry mixes with which to make your own drink. They come naturally and artificially flavored and colored and already contain the sugar needed. All you do is add the water. When ready to drink, their sugar and calorie values are about the same as in the other drinks—10–12 percent sugar and about 15 calories per ounce of liquid—but the cost to the consumer is much less than the canned drinks: under one cent per ounce. (Prices will vary with the price of sugar.)

While vitamin C is added to most of these dry products, the level of fortification varies from 15 percent RDA in a serving to 50 percent.

☐ SOFT DRINKS ☐

If any class of food product is associated with American life, it is soft drinks.

Our soft-drink technology is sought after by almost all other countries, including the U.S.S.R. and China. People introduced to our cola drinks apparently fall immediately in love with them. Other flavors follow in popularity.

When the discovery was made that water could be gasified, then sweetened and flavored, it set off a run on soda that knows no national boundaries or class distinctions. People seem to prefer soda to plain water — everywhere.

Most sodas contain 88–90 percent water, 10–13 percent sugar, and 120 calories per 8-ounce portion — just like the punches, drinks, and many fruit juices. However, they are *not* vitamin-C fortified and make no pretense of being any more than what they are: thirst quenchers, often artificially flavored and colored.

Unlike the canned fruit juices and drinks, sodas are not pasteurized and depend on their sugar content and acidity for preservation.

The diet sodas contain saccharin instead of sugar, and in some cases sodium benzoate is added as a preservative, since the sugar is no longer present to perform this function.

The addition of sodium benzoate is not a matter of necessity but of manufacturer's judgment. Some manufacturers believe that benzoate is required to keep the product fresh until consumption, others do not.

☐ BOTTLED WATERS ☐

There are a variety of bottled waters available, carbonated and still, imported and domestic. The consumption of these products has grown enormously in the United States, although bottled water has been used widely in Europe for a long time.

Its use in Europe was prompted mainly by inferior water collection and distribution systems in some countries. During the 1960s in the United States, when tremendous public concern was built up about the purity of our water, bottled-water sales jumped and have continued to grow, as people felt they were avoiding either bad-tasting or "impure" tap water. Also, home water "purification" devices have enjoyed a new popularity. In all but a very few highly publicized cases, concern about impurity is not well founded.

Another factor responsible for the explosive growth of bottled water sales is pure status and fashion appeal. Penny seltzer has yielded to Perrier, at very far from the original price of a penny.

The labels of the bottled waters sometimes show a mineral-salt analysis and thus imply, without stating, some sort of health benefit. The reason no health benefit is directly claimed is because such a claim would be illegal. There is no health benefit. The truth is that people have "heard" that such waters are "good for you."

These products do supply the pleasing sensation of carbonation. I always keep a bottle of cold carbonated water on hand, and when I want a drink other than plain water, I'll drink the carbonated water straight, or mix it with refrigerated orange juice, half and half.

Amazingly, if you keep water in a bottle in your refrigerator, you'll find that you will drink that more and soda less.

Perrier, of course, has elevated the drinking of carbonated water to high fashion, and it is entirely respectable to order Perrier or some other imported bottled water rather than an alcoholic drink or a soft drink.

The health inferences of the bottled-water manufacturers are probably warranted when you think about what you *don't* drink instead!

□ INGREDIENTS USED IN THIRST QUENCHERS □

■ WATER

Not much comment about this, except you will sometimes find the ingredient "filtered water" listed. This means little or nothing. The alternative to filtered water is not automatically dirty water.

■ THE FRUIT ACIDS

Citric acid (GRAS), malic acid (GRAS), tartaric acid (GRAS), fumaric acid (additive) are ingredients commonly found on drink labels. They are all found naturally in fruits. If natural fruit juices were labeled to show their ingredients, they would all show one or more of these acids.

The acids used may be either synthetic in origin or extracted from a fruit. That is unimportant; citric acid is citric acid no matter where it comes from, as long as it is pure citric acid. Sodium citrate (GRAS) is used to control acidity in some drinks.

Each of the fruit acids is used to provide a different tart character: citric for a citrus-fruit tartness, malic for apple type, tartaric for grape. Often a blend is used in an effort to provide an attractive taste.

The more fruit juice is diluted, as in the "drinks," the greater the necessity for adding fruit acids to make up for the watered-down taste that otherwise results. This watering-down process creates a need for many more additives as well.

■ SWEETENERS

Sugar may come from sugar cane or sugar beets — there is no difference. Corn syrup is made from cornstarch that is treated with enzymes. It consists mainly of dextrose. Dextrose has traditionally been cheaper than sugar and is only about half as sweet, ounce for ounce.

Corn sweetener is a product developed in the 1970s. It is made from cornstarch with a new enzyme system that converts the syrup into a mixture of dextrose and fructose, with control over the amount of fructose formed. This new corn sweetener has the same sweetness as sugar, is cheaper to make, and therefore has rapidly replaced sugar in many products where the sweetener is dissolved in water. Corn sweetener is not available dry, although fructose is now sold in health-food stores at outrageous prices unjustified by the cost.

■ GUMS AND THICKENERS

Dextrin, gum arabic, xanthan gum, ester gum, cellulose gum, sodium carboxymethyl cellulose, modified cornstarch are terms you run into often on drink labels. These materials are present to regulate consistency, to thicken, so that the drink does not "feel" watery.

Dextrin, gum arabic, and sodium carboxymethyl cellulose are GRAS. Xanthan gum and modified cornstarch are food additives, not GRAS.

Dextrin is a compound made from the starch found in corn and other cereals.

Gum arabic is obtained from the sap of the acacia tree and has been used for centuries as a thickener in medicines and foods.

Xanthan gum was developed by the U.S. Department of Agriculture in an effort to find new uses for products of corn. The material is made by the fermentation of dextrose (corn sugar) using a special strain of bacteria. Manufacturers of food products use xanthan gum because a small amount of it goes a long way (it's a powerful thickener), and it is not very sensitive to changes in temperature and acidity of the food. It has not been around long enough to be GRAS but has been tested sufficiently to be permitted as a food additive.

Cellulose gum and *sodium carboxymethyl cellulose* and *ester gum* are the same thing. Cellulose gum and ester gum are the common names, and the other is the accurate chemical name. The gum has been in use since 1945 and is common not only in processed foods but also in drugs, paper, textiles, and paints. Although its unlikely source is chemically treated cellulose from wood pulp or cotton, it is GRAS.

Modified cornstarch: The law requires labeling as "modified" any food starch that has been treated chemically. The methods of modifying are many: acid treatment, bleaching, oxidizing, etherifying (combining with an ether), esterifying (combining with an ester), or any combination. Acid treatment, bleaching, and oxidizing have far more history behind them than the other treatments. When flour is bleached, we both oxidize and bleach the starch in it. This raises an interesting problem. Bleached flour doesn't even come under the food additives amendment to the law, yet a food additive (modified starch) is produced within the flour as one of the components of flour when it is bleached with chlorine. Either chlorine-bleached starch should not be considered a food additive or bleached flour should be.

The term *modified starch* worries me. I would like to know how the starch is modified, and I would be less likely to select a food with esterified starch (and the possible complex compounds formed, about which we may not know enough) than one with oxidized starch, since I have been eating large quantities of oxidized starch for many years in the form of bleached or aged flour products.

■ CARAMEL AND ARTIFICIAL COLOR

All colored sodas and drinks contain either caramel or artificial color. Caramel color is the one present in cola drinks, root beer, and other brown drinks. The oranges, reds, and greens are all dyes that have been tested for safety. Caramel color is made by heating sugar. On rare occasions, oleoresin turmeric, a natural color, is used. It is GRAS.

■ ARTIFICIAL FLAVORS

Sodas that are artificially flavored carry the phrase ARTIFICIALLY FLAVORED in large letters on the front of the label.

Any one of more than 700 materials permitted as food additives — not GRAS — might be present. The quantities present are likely to be infinitesimal. The materials are usually derivatives of natural flavoring materials; for example, vanillin is naturally found in the vanilla bean and is the most important component that provides vanilla flavor. Synthetic vanillin is GRAS. In addition, vanillin acetate — a synthetic material that is more stable than vanillin in processed foods — is permitted as a food additive.

So, cream soda — 100 percent artificially flavored — could contain vanillin as well as vanillin acetate. The term "artificial" does not distinguish between GRAS and food-additive status.

There are a number of artificial flavors that are GRAS: synthetic compounds that duplicate materials found in nature, such as vanillin, mentioned above, or cinnamaldehyde (the substance primarily responsible for the flavor of cinnamon), or methyl anthranilate, responsible for the dominant note in grape flavor.

These materials, though they are identical to the same substances found in the natural plants, are artificial flavors. There is no way for consumers to tell whether the artificial flavor in their drink occurs naturally in plants or not. It is irrelevant to safety, anyway. If you are nervous about the safety of artificially flavored drinks, you should also be nervous about the safety of naturally flavored drinks. There are some natural flavors that have been found to be dangerous, such as safrole, which we discussed in chapter 4. But, there is absolutely no evidence of any kind that indicates danger from either source of flavorings now permitted for use.

■ NATURAL FLAVORS

These are materials that are extracted from plants so that the flavor exists in very concentrated form and then can be used to flavor a lot of something else. The natural flavors used in drinks are virtually all GRAS.

Quinine is responsible for a bitter taste in soft drinks. It comes from Cinchona bark. It is an additive. The maximum amount allowed is minute: not more than 1 part in 10,000; not enough for any drug effect.

■ MISCELLANEOUS SUBSTANCES

Brominated oil is added to make a carbonated drink look more natural, which everybody knows it is not. The brominated oil produces a cloudy effect and distributes flavoring. Ordinary oil would do it, but it would float to the top and become unsightly. If the oil is brominated, it then has the same specific gravity as the soda and tends to hang in there without separating. It is not GRAS. Brominated oil is found in orange soda and is always declared on the label. It is only permitted on an interim basis, until further testing is completed. Why not eliminate it until retested? There are plenty of orange sodas without it.

Propylene glycol is a GRAS material. It is widely used to dissolve flavor ingredients and is introduced into the drink incidentally when the flavor is added. It is present in many foods naturally.

Tricalcium phosphate is found in the dry drink mixes. It is an anticaking agent used in quantities well under 1 percent, is GRAS, and is a natural body substance. An anticaking agent prevents a powder from clumping when it gets slightly damp (which happens when people put wet spoons into powders) or undergoes temperature changes. It keeps the product free-flowing and easy to disperse. It is a very useful, totally safe material.

BHA and BHT prevent oxidation of fatty materials. The essential oils of lemon, for example, can oxidize quite readily when dispersed in a dry powder, as in the dry drink mixes. The process can be delayed by the use of BHA and/or BHT. Butylated hydroxyanisole and butylated hydroxytoluene are the full names. The FDA has agreed that BHA and BHT are easier names to handle and permits these abbreviations to be used. Both are food additives and are not GRAS.

Caffeine is present in drinks for flavor and stimulant effect. Caffeine is a drug but is GRAS when used at a level of not more than 0.02 percent or 1 part in 5000. (Recently, it has been classified as GRAS (3), which means more tests are needed soon.) This would provide 40 milligrams in 12 ounces of cola drink. A cup of coffee provides 150 milligrams. A dose that might be used in some pharmaceuticals is 200 milligrams. Therefore, a 12-ounce cola drink is about one fifth of a stimulating dose.

Ascorbic acid is vitamin C and is GRAS.

■ SACCHARIN

I have saved this for last in the discussion of additives to drinks because the artificially sweetened, low-calorie soft drinks are the principal source of saccharin in our diets. At the time this book is being written, the fate of saccharin is in limbo. Major issues have been raised by the discovery that saccharin causes cancer in laboratory animals — not to a very great degree.

There are two principal questions that have been previously mentioned regarding the saccharin issue:

First: Would more deaths be caused by eliminating the use of saccharin because, it is claimed, people would gain weight and increase their tendency to heart attacks?

Second: In view of the Delaney Clause to the Food Additives Amendment of 1958, can any substance that is found to cause cancer in man or animal be used in food, no matter what its advantages? Many scientists think that the Delaney Clause does more harm than good and is unrealistic. They argue that many substances, including many foods, may cause cancer in man or animals at very low incidence levels, but merely haven't been tested yet, and urge that the judgments of reasonable safety be used, rather than any special criterion for cancer.

At this time there is a renewed delay in the decision on saccharin. Every soda label and package of any other food containing saccharin must carry the warning: USE OF THIS PRODUCT MAY BE HAZARDOUS TO YOUR HEALTH. THIS PRODUCT CONTAINS SACCHARIN WHICH HAS BEEN DETERMINED TO PRODUCE CANCER IN LABORATORY ANIMALS.

The odds are that saccharin would be permitted if the Delaney Clause were to be repealed, since the hazard level seems to be very low.

☐ READING THE LABELS ☐

Now that we've prepared ourselves to read drink labels, let's look at one for each type of drink.

■ A NONCARBONATED FRUIT DRINK

The ingredients are: "Water, sugar, and corn sweeteners, concentrated orange juice, fumaric, citric, and malic acids (provide tartness), concentrated tangerine juice, sodium citrate (controls acidity), natural flavors, vitamin C, artificial colors. Contains 10% fruit juices."

This drink is fortified with enough vitamin C to give 100 percent of the RDA in a 6-ounce serving.

While the label says "made with real fruit juices," only one-tenth is real fruit juice, as shown in the ingredients statement.

The ingredients statement shows artificial color but no artificial flavor, and explains the purpose of each ingredient, which you are now familiar with.

You are asked to refrigerate the product after opening because the drink would ferment if allowed to remain open at room temperature.

This drink was 3 percent cheaper than pure pineapple juice in the same size container in the same store on the same day, and 30 percent cheaper than apple juice.

■ A FROZEN CONCENTRATE FOR IMITATION ORANGE JUICE

Imitation fruit drinks are designed to provide low-cost drinks that replace true-fruit-containing beverages. They can be dry mixes or similar in use to concentrated juice. Let's look at a frozen product. Is it for people who want to be able to buy their imitation product in its most inconvenient form — frozen and canned — or who believe that the only way to get good-tasting orange juice is out of the freezer in a can, even if it is imitation?

This label is totally informative. The ingredients listed are: Water, sugar syrup, corn syrup, orange pulp, citric acid, tripotassium phosphate, modified cornstarch, cottonseed oil, potassium citrate, tricalcium phosphate, vitamin C, natural and artificial flavors, sodium carbomethoxylcellulose, artificial color including F, D and C number 5, thiamine hydrochloride (vitamin B_1), BHA (preservative).

It says "contains no juice," although it does contain orange pulp. The tripotassium phosphate is used to control acidity. This product has cottonseed oil in it (although from its position on the label, it must be a mere fraction of a percent), which is added to make the product look cloudier than the orange pulp alone would do.

Quite surprisingly, this product is fortified with a small amount of vitamin B_1 (thiamine hydrochloride), and also tricalcium phosphate. The reason for such fortification escapes me. Fruit juices are not normal sources of these nutrients, but perhaps they are added to provide a sales claim. Ten percent of the RDA of thiamine and 4 percent of the RDA of calcium are provided.

Not surprising, and well justified, is the fortification with vitamin C so that one serving provides 150 percent of the RDA. We do expect to get our vitamin C from fruits or juices, whether they are real or imitation!

Potassium citrate is used instead of the usual sodium citrate to help control acidity. It is harmless. Our blood normally contains sodium and potassium citrates for the same reason the imitation orange juice does: to help control acidity.

Ethyl maltol is a flavor enhancer. Maltol, a chemical first cousin, is also used for this purpose. The substance does not have much of a flavor of its own, but it apparently increases the sensitivity of the taste buds, or causes more saliva to flow, thus increasing the perceived flavor. In fruit products, maltol increases the fresh taste, especially for fruits that have been heated in processing. As little as 75 parts per million is required. Another use may be to reduce the amount of fruit or flavor needed in a formula by making the flavor more "efficient," thus reducing the cost to the manufacturer. Maltol is a food additive.

■ A CARBONATED BEVERAGE (SODA POP)

There is a tremendous variety available. You can buy products with or without sodium benzoate, with or without artificial flavors, with or without artificial colors.

Two ingredients turn up that have not been mentioned in this chapter: erythorbic acid and EDTA. Erythorbic acid is a close relative of vitamin C and is GRAS. It is an antioxidant, that is, it tends to retard spoilage due to the action of oxygen. EDTA is the abbreviation for ethylenediaminetetraacetic acid. It is classified as a food additive. The purpose of EDTA is to capture traces of metallic material and to prevent their destructive action on flavor.

It is interesting to note that one product contains propylene glycol, yellow dye (F, D and C number 5), brominated oil, sodium benzoate, erythorbic acid, and EDTA. I cannot conceive what would require such a load of materials beyond the carbonated water, sugar, flavoring, and citric acid required to make soda. Not that there is any reason to suspect danger, but why would anyone want to buy such a product when less complicated and more wholesome-sounding products are available?

When evidence of danger is absent, this does not mean that there is proof of safety. It may mean that there has not yet been enough investigation, or that the substance has not been in use long enough for evidence of danger to develop.

There is a continuing examination of substances added to foods, almost always resulting in new evidence about some materials previously considered safe, and then found to be doubtful. This could even be because of a marked increase in the use of a substance. Therefore, to state it once again, my philosophy is to avoid excessive consumption of anything, and to stay away from strange materials that are not necessary. I can manage without saccharin, hence I avoid it — not absolutely, but certainly most of the time.

In other words, look for the simplest product that is *also* most enjoyable, and be wary of products with barely enough room on the label for all the added materials, when there is a competing product with fewer.

Nutritionally, sweetened carbonated beverages are a source of nothing but calories. Psychologically, they are clearly a source of pleasant thirst-quenching.

Whatever it is, the attraction of such drinks is universal. All over the world, regardless of the economic system in operation, there is unity on one subject: the desirability of soft drinks, especially the cola drinks.

■ CARBONATED WATER

No calories. No saccharin. The chemicals you see listed are found in natural mineral waters, which means nothing. They just give the water a slightly mineral taste. All are GRAS.

□ WHAT TO DO □

1. Remember, water is what quenches thirst. Drink it more often. Sodas are frequently just a way for us to get at a *cold* drink.
2. Save money and calories by mixing canned, frozen, or refrigerated fruit juices with water or club soda, half and half.
3. If you must drink sodas, moderation is in order — not quarts daily — and if you are overweight, artificially sweetened sodas are probably better than high-calorie sodas.

□ SUBSTANCES ADDED TO THIRST QUENCHERS □

MATERIAL ADDED	LEGAL STATUS (and GRAS class)	PURPOSE
Artificial color	Additive	Coloring
Ascorbic acid	GRAS (1)	Nutrition
BHA	GRAS (3)*	Antioxidant
BHT	GRAS (3)*	Antioxidant
Brominated oil	Interim Additive	Makes liquid cloudy, flavor carrier
Caffeine	GRAS (3)	Flavoring and stimulant
Caramel color	GRAS (1)	Coloring
Cellulose gum	GRAS (1)	Thickener
Cinnamaldehyde	GRAS (not classified	Flavor
Citric acid	GRAS (1)	Acidifier
Dextrin	GRAS (1)	Thickener
EDTA	Additive	Preservative (sequestrant)
Erythorbic acid	GRAS (1)	Antioxidant
Ester gum (see cellulose gum)		
Ethyl maltol	Additive	Flavor enhancer
Fumaric acid	Additive	Acidifier
Gum arabic	GRAS (2)	Thickener
Malic acid	GRAS (1)	Acidifier
Methyl anthranilate	GRAS (not classified)	Flavor
Modified cornstarch	Additive (1–5)†	Thickener
Oleoresin turmeric	GRAS (1)	Coloring
Potassium citrate	GRAS (1)	Acidity control
Propylene glycol	GRAS (1)	Dissolves flavors
Saccharin	Interim Approval	Artificial sweetener
Salt	GRAS (4)	Seasoning
Sodium benzoate	GRAS (1)	Preservative
Sodium carboxymethyl cellulose (cellulose gum)	GRAS (1)	Thickener

*GRAS (3) in fats and oils. Otherwise an additive.

†Modified food starches are not now specifically identified on the label. They may be anything from class 1 to 5, depending on which starch is used.

MATERIAL ADDED	LEGAL STATUS *(and GRAS class)*	PURPOSE
Sugar	GRAS (2)	Taste
Tartaric acid	GRAS (1)	Acidifier
Thiamine hydrochloride	GRAS (1)	Nutrition
Tricalcium phosphate	GRAS (1)	Free-flowing agent
Vanillin acetate	Additive	Flavoring
Xanthan gum	Additive	Thickener

Chapter 15

FROZEN DESSERTS

Ice Cream, Ice Milk, Sherbet, Water Ice

These frozen desserts differ from each other primarily in two aspects:

1. The amount of milk fat they contain;
2. The amount of air they contain.

The chart below shows the typical formulas used (adapted from *Encyclopedia of Food Technology*).

□ **ICE CREAM** □

	PREMIUM (%)	AVERAGE (%)	ICE MILK (%)	SHERBET (%)	ICE (%)	SOFT-SERVE (%)
Milk fat	16.0	10.5	3.0	1.5	—	6.0
Mild solids not fat	9.0	11.0	12.0	3.5	—	12.0
Sucrose	16.0	12.5	12.0	19.0	23.0	9.0
Corn syrup solids	—	5.5	7.0	9.0	7.0	6.0
Stabilizer	0.1	0.3	0.3	0.5	0.3	0.3
Emulsifier	—	0.1	0.15	—	—	0.2
TOTAL SOLIDS	41.1	39.9	34.45	33.5	30.3	33.5
Pounds/Gallon of mix	9.17	9.36	9.46	9.48	9.4	9.30

Draw from Freezer (After Freezing)

	PREMIUM (%)	AVERAGE (%)	ICE MILK (%)	SHERBET (%)	ICE (%)	SOFT-SERVE (%)
Overrun (%)	65–70	95–100	90–95	50	10	40
Approx lb/gal finished product	5.4	4.6	4.8	6.25	8.5	6.5

There is also a difference in the amount of sugar, especially between sherbet and water ice on one hand and ice cream and ice milk on the other. Nearly twice as much sugar is used in water ice as in ice cream.

Let's compare calories (per half cup):

ICE CREAM		ICE MILK	SHERBET	WATER ICE
Premium	*Average*			
160	130	100	120	170

The question that naturally arises is: How come the lower-fat product, water ice, has more calories than the highest-fat product, premium ice cream?

The answer lies in that word "overrun." None of these products would be pleasant to eat unless they were made with air blended into them. The texture of ice cream depends largely on the air that is beaten in. Without the air, it would be like a block of ice — more like a water-ice stick, which is not mixed while being frozen.

As you can see from the first chart, the "overrun" varies quite a bit. The percent of overrun expresses, in an indirect way, the amount of air in the product. When one gallon of liquid ice-cream mix with air injected yields two gallons of ice cream, it has 100 percent overrun, or 50 percent air by volume (not by weight).

Since water ice has only 10 percent overrun (or 5 percent air by volume), and premium ice cream has 70 percent overrun equal to 35 percent air by volume, ½ cup of ice cream *weighs* much less than ½ cup of water ice, so that in fact you eat much less in ½ cup of ice cream than you do in ½ cup of water ice.

Unfortunately, there is no way for you to tell from the label what the overrun is and therefore no way for you to compare value when you shop. All you can do is go by your taste. It would be helpful for consumers to know how much their quart of ice cream *weighs*. It might turn out that the seemingly most expensive product really is not!

Frozen desserts are the *only* foods sold that contain lots of air and that do not require the weight of the product to be listed on the label. Many baked foods contain air, but they are sold by weight. One wonders how this discrepancy ever came about. The dairy lobby must have made a pretty strong argument that people eat ice cream by the quart, not the pound, and therefore they need to know the volume, not the weight.

There are government standards for these desserts that do limit the amount of air permitted. This is accomplished by requiring that each gallon weigh *at least* a certain amount:

Ice cream and ice milk:	*not less than* 4.5 pounds per gallon
Sherbet:	*not less than* 6 pounds per gallon

There is no minimum for water ice stated since it is very difficult to include air in it anyway.

Most frozen dessert products now state the nutritional analysis per serving, so that we can have some control at least over our calorie, fat, protein, and carbohydrate consumption, if we use the information.

A peculiarity of our frozen-dessert consumption, however, is that most of it is not at home, most of it is not purchased in packages in stores. Therefore, we rarely know what we're getting, since there is no label at all to tell us. I've tried to read the label on a water-ice-confection stick purchased from a street vendor, and it is pretty difficult because it is crinkled or wet, and the print is tiny. (I've yet to see a ten-year-old kid *reading* his ice-cream package as he tears it off in order to get at the goody beneath.)

When you buy ice cream that has been packed into the container by hand, a lot of the air is squeezed out of it, and you really don't know what the overrun is. You can safely figure, however, that a scoop of ice cream from a hand-packed package will weigh 30 percent more than a scoop from a machine-filled package. (The air percentage drops from 45–50 percent down to 20 percent.)

■ MORE ABOUT THE GOVERNMENT STANDARDS

Ice cream is one of those products that can be artificially colored without the label saying so, although this is soon to be changed. Labels on ice milk, sherbet, or water ice, however, must say so if the product is artificially colored. I think this discrepancy originated with the fact that butter can be artificially colored without declaration, and butter is often used as the source of milk fat in ice cream, so it would be "inconsistent" and, incidentally, contrary to the interest of the dairy industry and lobby, to declare artificial color in ice cream. The FDA *recommends* that color be declared when it is used, and at least some manufacturers do this.

Whereas most foods must declare every ingredient, this is not so for ice cream, ice milk, or sherbet, because of the present standards for those products. The standards permit over twenty-four different sources of milk fat and milk solids. Those of us who harbor the quaint notion that all ice cream is made with milk and cream are due for a surprise. Sure, some ice cream is — the highest quality is made with fresh milk and cream — but *most* is not.

Dry milk, butter oil, and milk solids may be used. While none of these is synthetic or harmful, somehow ice cream made from dried cheese whey, butter oil, and dried buttermilk, all with water added, might not taste the same as when made from fresh cream and fresh skim milk. And the label doesn't have to tell you which substances are used! The government standards specifically say that any of those twenty-four sources can be declared as milk fat and nonfat milk. This is patently unfair to the consumer and misleads the buyer as to the value of the product. The argument of some manufacturers in support of this is lame: "As long as the amount of required fat and milk solids is present, and as long as the ingredients used are wholesome, why should exact declaration be required, especially since sometimes we have to shift from one source of milk fat or milk solids to another?"

□ ADDITIVES AND FROZEN DESSERTS □

Quite a few materials that are not "native" to ice cream are used in many brands of frozen desserts. With the industrialization of the frozen-dessert business, new demands were placed on formulation to help the product survive prolonged storage and once-a-week shopping. Just stop to think what ice cream must go through before it ends up at your dinner table, or in an ice-cream cone in your hand.

Since it is sold by volume, not weight, precise control of the amount of air desired is necessary. This is achieved by machinery and added materials, such as emulsifiers and gums.

Since it may be stored for a long time in the hardening room in a warehouse, then in the back-room freezer in a store, it must be protected against ice crystals growing and destroying the creaminess, making the ice cream feel "sandy" or "grainy." This protection requires added materials.

Since so many store freezers don't keep things frozen well, the dessert must be protected against melting and running out of the package in the freezer, and it must survive the trip home and back into the freezer, and maybe a few weeks in the home freezer at too high a temperature without being completely sealed.

Or how about the product that is put into an ice-cream truck with dry ice that freezes the stuff hard as a rock, then is sold to a nine-year-old boy in his clean outfit in the hot summer sun! If the product melts faster than the child can eat it, the result is one dripping child and a distressed mother. Or, even at home, consider the problem of getting the ice cream served to everybody for dessert before the first scoopful served melts. Many materials, primarily the gums, are used to control "melt-down rate."

Up to now, we haven't mentioned selling price. A 16-percent-butterfat ice cream, with 40 percent overrun, tolerates a lot more mistreatment than a 10-percent-butterfat product with 100 percent overrun. It can also cost *four times as much.* The mass market for ice cream is not for the most expensive products, which are made even more expensive by the use of all-natural flavors. Therefore, some ingredients are added to provide a satisfactory product at a price within the reach of most people.

While we are on the subject of expensive ice cream, a word or two about French ice cream (also called frozen custard or French custard ice cream). This is the same as regular ice cream, except that it must contain not less than 1.4 percent egg yolk *solids* by weight. This is equivalent to using 2.8 percent egg yolks. It is the egg yolks that are responsible for the yellow color. Note: there is *no* yellow vanilla ice cream made without eggs unless it is artificially colored.

■ VANILLA ICE CREAM, VANILLA-FLAVORED ICE CREAM, ARTIFICIALLY FLAVORED VANILLA ICE CREAM

What's the difference?

You will find all of these products in the freezer. The law says:

1. If *no* artificial flavor is used, the product may be called *vanilla ice cream.*
2. If artificial flavor is used along with natural flavor, but most of the flavor comes from the natural flavor, it must be called *vanilla-flavored ice cream.*
3. If there is artificial flavor used, even with natural flavor, but most of the flavor comes from the artificial flavor, it must be called *artificially flavored vanilla ice cream.*

This also applies to the labeling of ice milk, but not to sherbet. In the case of sherbet, if artificial flavoring is used at all, the label must clearly state that it is artificially flavored! I don't know why the standards are inconsistent.

With this background, let's go on to some labels.

Product A

All-Natural Ice Cream: Natural Vanilla Flavor

INGREDIENTS: Milk, cream, sugar, natural flavor
PRICE: 86¢ per pint

Product B

Vanilla-Flavored Ice Cream: Artificial Vanilla Flavor Added

INGREDIENTS: Milk and nonfat milk, corn syrup, sugar, whey, mono- and diglycerides, natural and artificial flavor, carob-bean gum, cellulose gum, guar gum, polysorbate 80, carrageenan, artificial color.
PRICE: 76¢ per pint

Product C

Vanilla Natural-Flavor Ice Cream

INGREDIENTS: Cream, milk, nonfat milk, sugar, corn sweetener, natural vanilla flavor, carob-bean gum,* guar gum,* lecithin.* No color added.
PRICE: $1.39 per pint

Product D

Vanilla Ice Cream

INGREDIENTS: Fresh cream, fresh skim milk, cane sugar, yolk of egg, natural vanilla
PRICE: $1.55 per pint

The most expensive product is twice the price of the cheapest product (100 percent more expensive), but consider: the pint of Product B weighs 283 grams (10 ounces), while the pint of Product D weighs 428 grams (15 ounces). Therefore, the price on a weight-comparison basis is: cheapest — 7.6¢ per ounce; most expensive — 10.3¢ per ounce. Product D is still more expensive, but "only" 33 percent more.

Let's look at each of the ingredients in Product B, our 76¢ bargain. The milk and nonfat milk could be any one of the

*Natural ingredients to maintain quality.

twenty-four ingredients permitted, we can't tell which. Mono- and diglycerides are added to increase the air-holding capacity, and to emulsify the fat into the tiniest-possible globules, and to help make lower-fat ice cream taste like richer, higher-fat ice cream (not a bad result). The gums are added to stabilize, that is, to retard the rate at which the ice cream melts, and to retard ice-crystal growth during the storage life of the ice cream, thus keeping it smoother. The manufacturer of this product thinks four gums are needed to do this, since carrageenan is also a gum, extracted from seaweed. Polysorbate 80 is also an emulsifier; however, unlike the mono- and diglycerides, it is not GRAS but a food additive, and not more than 0.1 percent is permitted in ice cream. The artificial color is voluntarily declared. Corn syrup is used to increase solids content without making the product too sweet and also to retard large-sugar-crystal formation during storage of the ice cream, which would make it taste "sandy."

In summary, all of the extra materials are there to make the product hold more air to begin with, to seem richer than it is, and to lengthen its storage life.

Somehow, Product D doesn't need all this and can get a very high price for the product. There is simply no doubt whatever about the ingredients used in this ice cream: they couldn't be simpler. Of course, Product D doesn't contain as much air in the first place, so there is not any great problem keeping the relatively small amount in.

But there are a few mysteries: Product A weighs just about the same as Product B but contains no emulsifiers, no gums, no corn syrup. How can this be? Especially since the same manufacturer makes Products A and B! And why does Product C need all those gums? And the lecithin (the natural emulsifier extracted from soybean oil)? I do not know the answers to these questions.

Which to buy?

For economy and best taste, I would buy the 86¢ product (A). For real enjoyment, I would buy Product D, if the money was not important. The one thing I would not buy is Product C: high price plus all the unnecessary additives.

What about calories and the nutritional label? Only A and B

carry a nutritional label. No matter how you pick and choose, if it's ice cream you are going to eat, ½ cup will not provide less than 150 calories and could provide as much as 200 calories. It is a safe rule that the more expensive the ice cream is, the higher the calories will be and the better it will taste. I'd rather eat the best less often, conserving money and calories and enjoying it the most.

■ ICE MILK

If you want to save calories and money, this is the product to buy. In addition to high overrun, ice milk contains less than half the fat of ice cream, and ½ cup will supply only 100 calories. Unlike ice cream, it is quite difficult to make creamy-tasting, rich-seeming ice milk without the use of gums and emulsifiers. This is a product that I think fully justifies their use. The chances are that there simply would be no ice milk, as an alternative to ice cream, without the use of these materials.

Virtually all of the brands of ice milk that I looked at contained at least three gums, at least two emulsifiers, and all contained only 2 grams of fat in a ½-cup serving compared with ice cream's 7, 8, or even 10 grams of fat.

The protein contribution of ice cream and ice milk is about the same: 3 grams in a portion, about 5 percent of the RDA, but the protein is of highest quality.

■ SHERBET

The most important thing to remember about sherbet is that it is rather dense, low in fat, and high in sugar. The chart below shows differences in formula between sherbet and other frozen dessert products.

■ ICE POPS

These are water-ice confections, frozen without stirring, and usually consist of sugar, water, natural and artificial flavors, colors, and gums. A 3-ounce pop provides 20 grams of sugar (5 teaspoons).

Natural flavors are beginning to be used more widely as a result of consumer pressure. No matter how natural the product claims to be, or is, the sugar contribution is high.

■ SOFT SERVE

This product is halfway between ice cream and ice milk in fat content, with low overrun. It also contains less sugar, and isn't as sweet as ice cream or ice milk. Because it is served without freezing it hard, it is extremely creamy tasting. Since soft serve is only available at the food-service level and not at home, you are unable to tell what's in it — whether it is naturally or artificially flavored or colored. I don't think it would be terribly burdensome to require a display of ingredients used in the soft-serve products, in the form of a simple sign at the point of sale.

□ SUBSTANCES ADDED TO FROZEN DESSERTS □

Material Added	Legal Status *(and GRAS class)*	Purpose
Artificial flavors	Can't tell*	Intensifies flavor
Carob-bean gum (same as locust-bean gum)	GRAS (2)	Prevents rapid melting
Carrageenan	Additive (3)	Thickener and prevents rapid melting
Cellulose gum	GRAS (1)	Prevents rapid melting
Guar gum	GRAS (2)	Prevents rapid melting
Lecithin	GRAS (1)	Emulsifier
Locust-bean gum (see Carob-bean gum)		
Mono- and diglycerides	GRAS (1)	Emulsifiers
Polysorbate 80	Additive	Emulsifier

*Unless the flavor is identified exactly, it is not possible to know whether it is GRAS or an additive.

□ NUTRITIONAL SUMMARY □

½-Cup Serving

Product	% Overrun	% Air	Fat (*grams*)	Sugar (*grams*)	Calories	Serving Weight/½-cup (*grams*)
Ice Cream	100	50	6.5	12	140	65
Ice Milk	95	47.5	2	13	100	68
Sherbet	50	25	1	25	124	88
Ices	10	5	0	36	172	120
Soft-Serve Ice Cream	40	10	5	14	132	92

Chapter 16

CANDY

Candy is mainly sugar. Nutritionally, candy supplies calories and nothing else. Psychologically, candy supplies considerable pleasure. Dentally, candy supplies problems.

Virtually every authority on nutrition agrees that we eat too much sugar and that we should reduce our consumption from the 130 pounds per year per capita that we consume. We actually spend more money every year on candy than we do on poultry — one of the best sources of nutrition!

But, let's face it, people like sweets and will continue to eat them.

Caramels, mints, hard candy, fudge, cream-filled candies, fondant, gum and jelly candies, taffy — each with a different texture — are possible because of the properties of sugar: the way it reacts when it is boiled down, cooled, whipped, crystallized, or treated with enzymes.

As anyone who has made syrup knows, we can dissolve as much as 65 pounds of sugar in as little as 35 pounds of water. If this is boiled down to almost no water, after mixing in a bit of citric acid, lemon flavor, and yellow color, and cooling, we end up with hard lemon candy.

If we add a bit of milk and fat to the syrup and cook down to 15 percent water, it will be caramel. Depending on how rapidly we cool and whether we want to crystallize, we can make creamy fudge or grainy fudge.

To make creamy, mint-filled chocolates, we form discs of mint-flavored, semi-hardened sugar with a bit of invertase (an enzyme) in it; a few days after coating with chocolate, the center turns creamy from the enzyme action. Or, if we don't want the center creamy, we may omit the enzyme invertase and make a mint patty whose center remains firm.

Invertase is an enzyme that converts sugar to dextrose and levulose — invert sugar — which does not crystallize and in fact prevents regular sugar from crystallizing, when present in small amounts.

Confectioners add invert sugar or corn syrup deliberately to some confections to prevent crystallization, that is, to keep the texture smooth.

To summarize, the differences between candies are the result of how the sugar solution is handled: how much water is boiled off, and how it is cooled. The water content in candy (excluding chocolates and nut candies) varies from less than 1 percent to as much at 15 percent. Sugar content varies from 75 to 90 percent.

Some candies contain more fat: chocolate candies, caramel fudge.

Unless the candy contains nuts, the protein content is negligible, even though milk may have been added. From 0 to 7 percent protein is the range, and that is absolutely insignificant. Pure *milk* chocolate contains only 7 percent protein.

□ CHOCOLATE □

"Chocolate," "chocolate flavored," "cocoa flavored," "vegetable-fat coating" are terms we will find on labels of apparent chocolate products. Nutritionally, there is no important difference; but there are some differences in cost.

Chocolate, a standardized food, is made from cocoa beans — botanically named cacao beans. After it is roasted and the fibrous shell removed, the bean is ground as fine as possible, producing bitter chocolate or "chocolate liquor," as it is called. This is common baking chocolate. No sugar or flavor has been added. Chocolate liquor is very high in fat, and the fat, or cocoa butter, can be easily removed by pressing, or extracting with a solvent.

If enough fat is removed so that what is left is a powder instead of a paste, that powder is cocoa, and it can contain from as much as 23 percent fat to as little as 10 percent. The flavor and quality of the cocoa as well as the price is generally considered to be directly related to the amount of fat in it. More cocoa butter: higher price and quality.

Cocoa can be improved in flavor and darkened in color by treatment with an alkali (usually potassium carbonate). This process is called dutching. Dutched cocoa is not cocoa made in Holland but is cocoa treated in this manner. The process is completely harmless as far as anybody knows.

So, the three main substances we get from the cocoa bean are chocolate liquor, cocoa butter, and cocoa.

The cocoa butter is a very unusual fat. It is hard, not greasy, at temperatures just below body temperature. It melts very quickly, with no greasy feel, in the mouth. Because of this property, cocoa butter is very useful not only as a food ingredient, but also in cosmetics and pharmaceuticals as well.

Chocolate liquor is used as an ingredient, along with cocoa and cocoa butter, to make fudge, chocolate coatings, and chocolate bars. Sugar, milk, lecithin, and flavor are added to make milk chocolate. If the milk is left out, sweet or semisweet chocolate is the result, both of which are darker than milk chocolate.

Chocolate bars — whether milk or sweet — contain about 56 percent sugar. Lecithin — sometimes declared as an emulsifier — is added because of another interesting property of cocoa butter. When cocoa butter hardens, its color is yellowish white. When chocolate is exposed to high temperatures, the cocoa butter melts slightly and some of it migrates to the surface of the chocolate; when the chocolate cools, this cocoa butters forms an unsightly whitish film on the chocolate, looking just like mold, which it isn't. This white film is called "bloom" and has no effect on flavor, but it completely destroys the appetizing gloss of chocolate. Lecithin is added to slow down bloom formation.

Finally, the artificial flavor added to chocolate is either vanillin or ethyl vanillin. Vanillin is used because chocolate cannot tolerate any water or alcohol, which would be present in vanilla extract, without losing gloss. Since the vanillin is dry, this problem is avoided. Ethyl vanillin is a synthetic substance three times as strong as vanillin and some say with a slightly smoother flavor. Both vanillin and ethyl vanillin are GRAS.

Because of the high cost of chocolate, especially cocoa butter, many candies today are coated with vegetable-fat and cocoa coating, flavored artificially or with essential oil of chocolate (distilled from chocolate), since it is cocoa butter, rather than cocoa, that provides the aroma we associate with chocolate. While they do not have the same flavor and consistency as true chocolate, when they are used as coatings over other confections they can be very satisfactory — in some ways even more satisfactory than true chocolate, from a physical point of view. They do not bloom as readily, do not require the cooling and heating control within minute fractions of a degree that chocolate does, and are less brittle.

These coatings are more likely to be found on baked products than confections; however, the trend seems to be back to real chocolate because of superior flavor and texture.

□ OTHER CANDY INGREDIENTS □

Many of the ingredients found in candy are now familiar to you, but there are a few that are principally used in confectioneries and are not very common in other foods.

Confectioner's glaze is used to coat a confection to prevent the chocolate coating on it from melting on one's hands. It consists of shellac, which is obtained from an insect. It is the same source of shellac that is used in painting but is prepared so that it meets food purity standards, that is, it is not contaminated by any toxic materials.

Glycerine is added to keep moistness in the candy or as a solvent for a flavor. It is GRAS.

Petrolatum is petroleum jelly, like Vaseline. It is used as a coating film to make a gum candy shiny — candy polish, if you will.

Fruit pulp is what is left if you squeeze the juice out of a fruit. This is used as a filling in some hard candies, along with artificial flavors and colors to "replace" what was removed.

Cream of tartar and tartaric acid are used to provide a tart flavor and to help to invert the sugar to prevent crystallization. (Inversion consists of changing the disaccharide sucrose into its monosaccharide components, dextrose and levulose.) The acid is a natural component of grapes and other fruits. Cream of tartar is the potassium salt of tartaric acid. Both are present in wine; in fact, these materials are made from waste products of the wine industry. They are GRAS.

Dextrin is used to regulate the consistency of confections. It is not as gummy as starch. Corn sugar is made by breaking down cornstarch to its simplest building block: dextrose. Before that simplest stage is reached, though, a carbohydrate simpler than starch but more complicated than dextrose is formed: dextrin. It is GRAS.

Sorbitol and mannitol are made from corn and are slightly sweet. Unlike sugar and dextrose, they are absorbed and utilized more slowly and are frequently used as sugar "substitutes." They also serve to control the moistness and softness of a food instead of glycerin. They are GRAS. If more than 50 grams of mannitol are eaten daily, there is a possibility of a laxative effect. If it is present in a food that may be consumed in quantities of 20 grams per day, the food label must state "Excessive consumption may have a laxative effect."

□ NUT CONFECTIONS □

Nuts, mainly peanuts, turn up in a number of candies: chocolate mixed with peanuts or almonds, peanut brittle, and some chocolate-covered peanut-and-caramel bars.

One chocolate-and-peanut bar provides in a one-ounce serving 8 percent of the day's protein in 150 calories along with 4 percent of the riboflavin, 8 percent of the niacin, and 2 percent each of the thiamine, calcium, and iron. That's not bad; in fact, when I was a student during the Depression, I often ate this bar for lunch out of sheer economy — with satisfaction (and for a nickel). Not surprisingly, this product (Mr. Goodbar) does voluntarily provide the nutritional analysis on the label.

There is another chocolate-coated peanut-butter candy in the shape of little cups (Reese's Peanut Butter Cups) with a very similar analysis in 210 calories.

Peanut brittle is over 80 percent sugar and is only 5.7 percent protein (it would take over 3 ounces to get 7 percent of the RDA for protein — and more than 400 calories).

Analysis of the other peanut candies with chocolate fudge and caramel are not stated on the label, but they contain enough peanuts so that if you really have to eat some candy, the calories won't be totally empty.

□ CHEWING GUM □

The amount of food or calories ingested by chewing gum is small enough to be ignored, unless you chew a pack a day. The main problem is the sugar contact with the teeth — hence, sugarless gums.

Chewing gum base may be one of more than 20 substances, some extracted from trees or plants, others synthetic, including butadiene — styrene rubber!

Softeners may be one of thirteen materials, including "glycerol ester of partially hydrogenated gum or wood rosin." That statement alone is bigger than the stick of gum! However, in time, the new FDA labeling objectives should lead to some indication on the label of what the mysterious gum base is and where the softener comes from.

These substances are food additives, not GRAS.

□ CONCLUSION □

Eat candy moderately and pick the ones you like the most. There are no nutritional bombs, positive or negative, to be discovered at the candy counter. If you want to soothe your conscience, pick nut candies. Like so many other foods, most are not nutritionally labeled. New FDA policy will urge mandatory nutritional labeling and removal of added vitamins and minerals in candy so that their presence does not encourage consumption of candy at the expense of other, more nutritious foods.

□ SUGAR AND FAT □

(Grams per Ounce)

	SUGAR	FAT
Hard Candy	27	0
Gum or Jelly Candies	25	0
Jelly Beans	26	0
Caramels: Chocolate with Nuts	20	5
Chocolate Fudge	20	5
Peppermint Patties (Chocolate Covered)	23	3
Chocolate-Covered, Coconut Center	20	5
Peanut Brittle	23	3
Milk Chocolate	16	9
Sweet Chocolate	16	10

Use this as a guide for candies that are not nutritionally labeled. This applies to "natural" candies as well, unless otherwise labeled.

□ SUBSTANCES ADDED TO CANDY □

MATERIAL ADDED	LEGAL STATUS *(and GRAS class)*	PURPOSE
Carnauba wax	GRAS (5)	Provides high shine to chocolates
Cream of tartar	GRAS (1)	Acidity adjustment
Dextrin	GRAS (1)	Texture adjustment
Ethyl vanillin	GRAS (not classified)	Flavoring

Material Added	Legal Status (*and GRAS class*)	Purpose
Glycerine	GRAS (1)	Prevents drying
Glycerol ester of partially hydrogenated gum or wood rosin	Additive	Texture
Invertase	GRAS (not classified)	Liquefies sugar
Lecithin	GRAS (1)	Emulsifier
Mannitol	Interim approval	Sweetener, adjusts firmness of candy center
Petrolatum	Additive	Produces surface sheen
Potassium carbonate	GRAS (1)	Makes chocolate darker
Sorbitol	GRAS (1)	Prevents drying out
Sugar	GRAS (2)	Basis of candy
Tartaric acid	GRAS (1)	Acidity adjustment
Vanillin	GRAS (not classified)	Flavoring

Chapter 17

HOW TO USE THE INFORMATION IN THIS BOOK

□ NUTRITIONAL CONSIDERATIONS □

More important than any other factor is the need to meet the 1980 Dietary Guidelines. (See appendix 1 for the complete government text.) These guidelines, summarized, are:

1. Eat a variety of foods. Every day include foods from the following groups: a) fruits and vegetables; b) whole-grain and enriched breads, cereals, and grain products; c) milk, cheese, and yogurt; d) meat, poultry, and fish, eggs, dry peas and beans.

The reason for this is to assure you of obtaining the basic nutrients: protein, vitamins, minerals, and essential fatty acids.

The diet shown *(over)* is a balanced, low-calorie diet, with 40 percent of the calories coming from fat, and with virtually no sugar in it. It supplies foods from the four groups and leaves room for additional foods while remaining within the recommended calorie range of 2000–2500 per day. You can add foods with some sugar and little or no fat, thus cutting the fat contribution to the day's calories down to the 35-percent level. The basic diet provides 1415 calories, with 567 of those coming from fat. Scientists haven't been able to agree on exactly what percentage of our daily calories should be from fat, but they do agree that it should be lower than we now consume, and somewhere in the 30- to 40-percent range. The American Heart Association recommends that the fat calories you eliminate be replaced by complex carbohydrates (starches).

The diet shown here is special in that it supplies almost all of the necessary nutrients from foods in a relatively small number of calories. What I like about this diet is that it specifies amounts of food and provides plenty of room for variation and addition. This diet is referred to as a "pattern dietary" in *Nutrition in Health and Disease.*

Don't try to be too precise about selecting your diet. You won't succeed unless you weigh everything and have an encyclopedia near you at mealtimes.

Let me give an example of how you might use the pattern dietary to build up calories and get some of the foods you might enjoy, while reducing the percentage of fat calories. Please do *not* consider this to be a *recommended* diet. I don't recommend specific diets: there are plenty of outstanding diet books, in addition to the dietary guidelines, that do.

In this example, you can go from 1415 calories to 2000 calories, with ice cream, jam, and cake included — and reduce your fat calories while doing it! Here's how:

- Replace the whole milk with skim.
- In the "meat, fish, poultry, egg" category, choose chicken rather than red meat, and use 8 ounces, raw weight, rather than 4 ounces as specified.
- Add a cup of yogurt.
- Add 4 ounces of cottage cheese.
- Double the amount of raw vegetables and use some low-calorie salad dressing.
- Use 2 teaspoons of jam on your bread.
- Have a doughnut or some cookies or a piece of cake.
- If you insist on gorging yourself on desserts, add some ice cream.

□ EVALUATION OF A PATTERN DIETARY FOR

Food Group	Amt. (g.)	Household Measure	Energy (Kcal.)	Protein (g.)	Fat (g.)	Carbo-hydrate (g.)
Milk or equivalent[b]	488	2 c. (1 pint)	320	17	17	24
Meat, fish, poultry, or egg[c]	120	4 oz., cooked	376	30	31	..
Vegetables:						
Potato, cooked	100	1 medium	65	2	..	15
Deep green or yellow, cooked[d]	75	½ c.	21	2	..	6
Other, raw or cooked[e]	75	½ c.	45	2	..	10
Fruits:						
Citrus[f]	100	1 serving	44	1	..	10
Other[g]	100	1 serving	85	..	..	22
Bread, white, enriched	100	4 slices	270	9	3	50
Cereal, whole grain or enriched[h]	130 30	⅔ c. cooked or 1 oz. dry	89	3	1	18
Butter or margarine	14	1 tbsp.	100	..	11	..
TOTALS			**1415**	**66**	**63**	**155**
Compare with recommended allowances[j]						
Males (70 kg., 23–50 yrs. old)			2700	56	..	..
Females (58 kg., 23–50 yrs. old)			2000	46	..	..
% RDA (Approx.)				100+	..	..

[a]Calculations from Composition of Foods. Handbook No. 8 U.S. Department of Agriculture, rev. 1963.

[b]Milk equivalents means evaporated milk and dried milk in amounts equivalent to fluid milk in nutritive content; cheese, if water-soluble minerals and vitamins have not been lost in whey; and food items made with milk.

[c]Evaluation based on the use of 700 g. of beef (chuck, cooked), 200 g. of pork (medium fat, roasted), 200 g. of chicken (roaster, cooked, roasted), and 100 g. of fish (halibut, cooked, broiled) per 10-day period, and egg occasionally.

[d]Evaluation based on figures for cooked broccoli, carrots, spinach, and squash (all varieties).

ITS NUTRITIVE CONTENT[a] □

MINERALS				VITAMINS				
CALCIUM (mg.)	PHOSPHORUS (mg.)	MAGNESIUM (mg.)	IRON (mg.)	A (IU)	THIAMINE (mg.)	RIBOFLAVIN (mg.)	NIACIN (mg.)	ASCORBIC ACID (mg.)
576	452	63	.2	700	.16	.84	.3	5
13	212	104	3.3	280	.14	.23	6.1	..
6	48	22	.5	..	.09	.03	1.2	16
44	28	29	.9	4700	.05	.10	.5	25
16	41	18	.9	300	.08	.06	.6	12
18	16	12	.3	140	.06	.02	.3	43
10	21	16	.8	365	.03	.04	.5	4
84	97	20	2.5	..	.25	.21	2.4	..
12	95	34	.9	..	.08	.03	.7	..
3	2	2	..	460	..	..	..	..
782	**1012**	**335**	**10.3**	**6945**	**.94**	**1.56**	**12.6[i]**	**105**
800	800	350	10.0	5000	1.40	1.60	18	45
800	800	300	18.0	4000	1.00	1.20	13	45
98	126	96	55–100	150	65	100	75	200+

[e]Evaluation based on figures for raw tomatoes and lettuce, and cooked peas, beets, lima beans, and fresh corn.

[f]Evaluation based on figures for whole orange and grapefruit, and orange and grapefruit juices.

[g]Evaluation based on figures for banana, apple, unsweetened cooked prunes, and sweetened canned peaches.

[h]Evaluation based on figures for shredded wheat biscuit and oatmeal.

[i]The average diet in the United States, which contains a generous amount of protein, provides enough tryptophan to increase the niacin value by about one third.

[j]From the National Research Council Recommended Dietary Allowances, revised 1974.

Here's what happens to the calories:

	CALORIES	FAT CALORIES
Whole milk, 2 cups	320	153
Skim milk, 2 cups	180	0
Calorie saving	**140**	**153**
4 ounces red meat	376	279
8 ounces chicken	300	54
Calorie saving	**76**	**225**
TOTAL SAVING	**216**	**378**
Now add:		
1 cup yogurt	150	36
4 ounces cottage cheese	120	36
½ cup ice cream	140	56
Extra vegetables, with dressing	115	27
2 teaspoons jelly or jam	35	0
Doughnut, cake, or cookies (2 ounces)	240	108
Calories added	**800**	**263**
Less calories saved	**−216**	**−378**
NET CHANGE	**+584**	**−115**
Calories in basic diet	+1415	
TOTAL CALORIES	1999	

So, compared to the pattern diet, we have added 584 calories to 1415 for a total of 1999, and reduced fat calories from 567 to 452, thus making our fat calories only about 22.5 percent of the day's total!

Another way to use the dietary base is simply to eat more of what is shown, particularly of the cereals and fruits. The way to make room for the high-fat desserts would have to be by eliminating high-fat foods in the pattern diet, such as whole milk and higher fat meats.

If you doubled the amounts of bread, potato, and cereal shown in the diet, you would add 424 calories (with added complex carbohydrates), of which only 27 would be fat calories. You would end up with 1839 calories, 594 of them fat calories — only 32 percent of the total. You would also increase your vitamins and minerals.

What is rather clear is that it is not easy, for women, especially, to get enough iron in their diets, without the use of an iron supplement or iron-fortified foods.

The pattern diet can be used to build up your calorie needs, or, by comparing what you eat with it, to reduce or add to your present diet to meet the guidelines.

2. Maintain ideal weight. (See chart, page 256.) Avoid crash diets if you need to lose weight. Lose weight gradually: 1 to 2 pounds per week. To lose weight, reduce the higher-fat foods and high-sugar foods first, trim meats thoroughly, remove skin from chicken, replace meat with fish, drink low-fat milk, eat cottage cheese instead of the higher-fat cheeses, avoid cream, ice cream, excessive sugared soft drinks, rich cake and pastry foods, high-fat snacks, such as potato chips.

3. Avoid too much fat, saturated fat, and cholesterol. Sure, some scientists have said that there is no proof that high cholesterol causes heart disease, but *no* scientists have said that high cholesterol is desirable. Play the odds and follow this dietary guideline. The best advice seems to be to cut down on total fat, saturated fat, and cholesterol.

Following this recommendation will help you control your weight as well. Remember: 1 gram of fat provides 9 calories — more than twice as much as 1 gram of protein or carbohydrate.

To control your cholesterol level, avoid excessive use of eggs, butter, saturated margarines, and organ meats such as liver.

Use your new knowledge of label reading to pick the foods with less fat and less saturated fat. Buy the brands that tell you!

4. Eat foods with adequate starch and fiber. Our refined Western diets have led to an increase in bowel problems, whereas more primitive diets in "backward countries" show

□ SUGGESTED BODY WEIGHTS □

(Range of Acceptable Weight)

HEIGHT *(feet and inches)*	MEN *(pounds)*	WOMEN *(pounds)*
4′10″		92–119
4′11″		94–122
5′0″		96–125
5′1″		99–128
5′2″	112–141	102–131
5′3″	115–144	105–134
5′4″	118–148	108–138
5′5″	121–152	111–142
5′6″	124–156	114–146
5′7″	128–161	118–150
5′8″	132–166	122–154
5′9″	136–170	126–158
5′10″	140–174	130–163
5′11″	144–179	134–168
6′0″	148–184	138–173
6′1″	152–189	
6′2″	156–194	
6′3″	160–199	
6′4″	164–204	

Note: Height without shoes; weight without clothes.
Source: HEW conference on obesity, 1973.

fewer bowel problems. We need fiber to reduce problems of constipation, diverticulosis, and perhaps even cancer of the colon. Eat fruits, vegetables, whole-grain breads and cereals. (If you are unable to eat such foods, buy foods with fiber added to them — you'll find it on the label.) HEW and USDA say that there is no need to add fiber to foods that do not already contain it. You should be able to get enough by eating the recommended fruits, vegetables, and whole-grain products. But from a practical point of view, it might be easier to use fiber-added breads, of which there are many, especially if you do not like whole-grain bread.

Starch is needed as a calorie source in the form of complex carbohydrates.

5. Avoid too much sugar. Whatever that means. Everyone agrees that a film of sugar on the teeth results in cavities. HEW says, "Frequent snacks of sticky candy, or dates, or daylong use of soft drinks may be more harmful than adding sugar to your morning cup of coffee — at least as far as your teeth are concerned."

There is no guideline provided for how much less sugar we should eat than the 130 pounds per year per person we now consume. I suggest we use 100 pounds per year as a guideline — about a 25-percent reduction. Let's see where that takes us: 100 pounds per year equals 124 grams per day, as compared with the 161 grams per day in 130 pounds. This means reducing our intake of sucrose, glucose, maltose, dextrose, lactose, fructose, corn sweetener, corn syrup, and honey by 37 grams per day; to keep it simple, let's set a goal of reducing our sugar consumption by 40 grams per day.

This is equal to 160 calories per day. Since cutting 3500 calories from your diet will result in a loss of one pound, if all you did was simply eliminate this much sugar, without replacing it with any other food, you would lose one pound every 22 days or 17 pounds in a year, assuming your activity level was unchanged. What a simple, healthy, uncomplicated, gradual way to reach ideal body weight!

How much is 40 grams of sugar?

One twelve-ounce can of a soft drink contains 39 to 40 grams of sugar.
One two-ounce candy bar, almost any kind, contains 35 to 40 grams of sugar.
One two-ounce piece of cake, with icing, contains 20 to 30 grams of sugar.
One level measuring teaspoon contains 4 grams of sugar. A rounded normal teaspoon contain 8 grams of sugar.

Remember, any of the sweet thirst quenchers is 10 percent or more sugar (unless saccharin-sweetened). Look at the label and take 10 percent of the amount of the product to see how much sugar you will be consuming. (This is easy with the metric system. Just take 10 percent of the amount in liters, and that will be approximately correct in grams.)

Even orange juice is 10 percent sugar, but by cutting it half and half with water or soda water, you still have a great thirst quencher, with only half the sugar of full-strength juice. (Unfortunately you also cut your vitamin C intake in half this way.)

If you eat sweet baked foods unsugared and un-iced, you will save one-third of the sugar.

If you like canned fruits, discard the sweetened juice or syrup—or better still, buy unsweetened canned fruit. A heavy-syrup-packed fruit contains three times as much sugar as the water-packed fruit. Save 15 grams in a 3-ounce serving in water pack versus heavy syrup.

If you don't think you are consuming 161 grams of sugar per day, please stop to realize that there is sugar added to jams, jellies, candies, cookies, soft drinks, cakes, pies, some breakfast cereals, catsup, flavored milks, ice cream, water ice pops, sherbet, and drink mixes.

6. Avoid too much sodium. The guidelines do not state what too much is. The main purpose of this advice is to reduce the incidence of high blood pressure and stroke. Sodium is one major factor that affects blood pressure; obesity is another.

For those on a sodium-restricted diet, under the direction of a physician, the use of sodium information on the label is necessary. As a guide for the average person, however, restraint in the use of salt and salty foods — *especially high-fat salty foods* like potato chips, salted nuts, salty cheeses — is advised. Potato chips can contain as much as 300 milligrams of sodium per ounce. Other foods to be controlled are pretzels, soy sauce, herb salts, pickled foods, pickles, cured meats (hot dogs, bologna, salami, ham, bacon).

The estimated safe sodium intake ranges from as little as 450 milligrams per day for four-year-old children to a maximum of 3300 milligrams per day for an adult. The amount required is not exactly known. For example, in adults alone, the suggested range is from 1100 to 3300 milligrams per day — that's 3.3 grams.

Probably the best way to control this is at the salt shaker. A reasonable shake of salt on an egg, which I weighed a few dozen times, was 250 milligrams, equal to 100 milligrams of sodium. Every time you shake your salt shaker, you get 100 milligrams of sodium. It is useful to know that ¼ teaspoon of salt weighs 1500 milligrams and contains 600 milligrams of sodium.

There is no need to worry about obtaining sufficient sodium, only about consuming too much. Rather than trying to compute every milligram, the general advice to follow is to restrict the use of salt and salty foods.

7. If you drink alcohol, do so in moderation. Moderation means one or two drinks daily. This would be equivalent to 8 ounces of wine daily or 16 ounces of beer.

The caloric content of alcoholic beverages varies considerably: 80 calories in an ounce of gin, rum, vodka, or whiskey; 152 calories in a 12-ounce bottle of beer; 115 calories in a 4-ounce glass of wine.

□ SELECTING FOODS □

These general guidelines, with some ideas for meeting them, are still not enough for the person who wants to make the most of food-label information. You will need to invest some time in reading labels to develop your preferred shopping list of products. In general, you will be able to identify companies that:

1. Use all sorts of additives, while other companies do not — for the same type of product.
2. Provide simple, clear, understandable labels, with realistic portion sizes; instead of barely legible, white-on-white printed labels, with misleading portion or serving sizes to make the product appear to be better than it is.
3. Provide information voluntarily on the label, such as cereal companies that state the sugar content on cereal labels.
4. Provide nutritional analysis even for the high-calorie, low-protein and -vitamin products (as Hershey's now does on chocolate).

Select products by brand. While I believe that you should always read the label before deciding on a food, after a while you will be able to associate a company's philosophy with its products through its labels (Dannon yogurt: "no artificial anything," Pepperidge Farm cookies: "all-natural flavors") and then you can have more faith in the brand and relax a bit.

What seems to make the most sense is to pick one food category at a time, get familiar with the brands and labels, pick your brand — never forgetting that you should enjoy your food too — and when you have settled on the products you wish to use in that category, go on to the next category. For example: you might decide to read only milk labels on your next trip and decide that you want your children to drink whole milk and you want to drink low-fat milk. On your next trip, select another group, say cheeses, and make your decisions. Watch for new products in each category: more and more foods are being developed to assist consumers in reducing their fat, sugar, and cholesterol intakes.

(Occasionally, after you have used a product for a while, read the label again when you get home. The products are sometimes changed. Just recently my favorite brand of cottage cheese was changed to include new gums. I did not like the product, and I therefore did a new reading of all cottage-cheese labels and selected a new brand.)

Sometimes the same company will have different brands to appeal to different consumers. Kraft Foods owns Breyer's *and* Sealtest. Breyer's Ice Cream is all-natural; the Sealtest label is used for many artificially flavored and colored products. This comment is in no way meant to be derogatory: it is actually a help to the consumer to be able to associate a brand with a type of product.

Beware of nutritional claims without nutritional analysis. Words like "natural" and "whole" and "organic" are often masks for inferior, more expensive foods. Look for the backup for any claim, direct or implied, on the label. If the backup nutritional information is not on the label, ignore the claim.

Watch out for irrelevant claims, such as the "no preservatives" claim on products that never needed, nor ever contained, any, like "natural," high-priced peanut butter.

Read labels carefully. One brand of refrigerated orange juice says: Valencia blend. The word "blend" is less than half the size of "Valencia." In still smaller type, the label informs us that there is "at least 50 percent" Valencia orange juice in the product. Does anyone want to bet on how much *more* than 50 percent there might be?

Read the entire label, not only the ingredients statement. Take the *Case of the Diminishing Orange Juice* (also see chapter 14). In the same frozen food case, there were three kinds of orange drink. The first was 100 percent pure orange juice concentrate. No ingredients statement. No nutritional label. Just plain frozen concentrated orange juice.

The second was a product called "Orange Plus," and underneath the name, it clearly stated "contains 30 percent orange juice"—nothing misleading about it. But unless the label is read carefully, some people might think that the product is pure orange juice that has been fortified. The label clearly shows what is added to the orange juice, including water, vitamin C, and flavors.

The third product, "Awake," contains no juice, and says so. There is nothing misleading about this label. It is a product made of nonjuice ingredients and water, which is frozen in a can the same size as the other two products. I found all three in the same freezer case. The price of "Awake" is much lower than either the juice concentrate or "Orange Plus." It is sold as a substitute for more expensive orange juice, plainly and clearly.

So you see, you can buy pure concentrate at one price; concentrate plus water and other materials including vitamin C at a slightly lower price; and imitation concentrate at nearly half the price of the pure juice concentrate. But, if all you were to do is look at the can color, and not read the labels and prices as well, you might believe that all three were some kind of orange juice.

The manufacturer is supplying what he thinks people want and perhaps need. But you are the one who must read the label and make the choice for yourself.

Our food supply is probably as safe as any in the world. There is growing responsiveness to consumer awareness not only by the FDA but by food manufacturers as well. More and more packages voluntarily contain information not yet required by law but informative to the consumer.

Still, there are many things that need improvement, and some that I believe are outright wrong, such as interim permission to use brominated oil. In the case of saccharin, there is a reasonable basis for interim approval: to fight obesity and consequent health damage. But it is difficult to see the justification for the use of brominated oil.

We need continuing consumer awareness and pressure on processors to eliminate what is unnecessary. Food manufacturers care more about what their customers think than any other single thing. Unless they can please and keep their customers, they won't be in business very long. Tell them what you think. The name and address of the manufacturer is on the label. Write to the company and ask your question or make your comment. You will be amazed at the response; and if you don't like the answer, change brands.

Finally, I suggest that you select the foods you buy based primarily on nutrition and taste. Very often the foods with the fewest additive materials are those that taste best. Once that choice is made, you should be able to select the brands you want depending on how strongly you feel about the ingredients used.

There are plenty of choices — ranging from only GRAS ingredients to the use of many additives — in the same type of product.

Here's to your health through informed eating! Remember — you will be eating for the rest of your life.

BIBLIOGRAPHY
APPENDIXES
INDEX

BIBLIOGRAPHY

Books and Pamphlets

Aykroyd, W. R., and Doughty, Joyce. *Wheat in Human Nutrition.* FAO Nutritional Studies, No. 23. New York: Unipub, 1970.

Code of Federal Regulations, Number 21 (Food and Drugs). Washington: U.S. Government Printing Office, 1979, 1980.

Desrosier, Norman W., ed. *Elements of Food Technology.* Westport, Connecticut: AVI, 1977.

Dietary Allowances Committee and Food and Nutrition Board. *Recommended Dietary Allowances.* Washington: National Academy of Sciences, 1980.

Federal Food and Drug Act as Amended. HEW pub. no. (FDA) 79–1051. Washington: U.S. Government Printing Office, 1979.

Food Protection Committee. *Food Chemicals Codex.* Washington: National Academy of Sciences, 1972.

Food Safety Council. *Proposed System for Food Safety Assessment: A Comprehensive Report on the Issues of Food Ingredient Testing.* New ed. Elmsford, New York: Pergamon Press, 1978.

Furia, Thomas E., ed. *Handbook of Food Additives.* 2nd ed. 2 vols. Boca Raton, Florida: CRC Press, 1979, 1980.

Harper, Harold A., et al. *Review of Physiological Chemistry.* 17th rev. ed. Los Altos, California: Lange, 1979.

Heid, John L. and Joslyn, Maynard A. *Fundamentals of Food Processing Operations: Ingredients, Methods, and Packaging.* Westport, Connecticut: AVI, 1967.

Johnson, Arnold and Peterson, Martin, eds. *Encyclopedia of Food Technology.* Westport, Connecticut: AVI, 1974.

Margolius, Sidney. *Health Foods: Facts and Fakes.* New York: Walker, 1973.

Mazur, Abraham and Harrow, Benjamin. *Textbook of Biochemistry.* 10th ed. Philadelphia: Saunders, 1971.

Merrill, A. L., Watt, B. K., et al. *Composition of Foods* (Agricultural Handbook no. 8, available from P.O. Box 17873, Tucson, Arizona).

Mitchell, Helen S., et al. *Nutrition in Health and Disease.* 16th ed. Philadelphia: Lippincott, 1976.

Passmore, R., et al. *Handbook on Human Nutritional Requirements.* Albany, New York: World Health Organization, 1974.

Peterson, Martin S. and Johnson, Arnold H. *Encyclopedia of Food Science.* Westport, Connecticut: AVI, 1978.

Protein Requirements. FAO Nutritional Studies, No. 16. New York: Unipub, 1950.

Requirements of Laws and Regulations as Enforced by the U.S. Food and Drug Administration. HEW pub. no. (FDA) 79–1042. Washington: U.S. Government Printing Office, 1979.

Steele, F., and Bourne, A. *The Man/Food Equation.* New York: Academic Press, 1975.

Transactions. Publication of the American Society of Bakery Engineers. 1960–1979.

Appendix 1

NUTRITION AND YOUR HEALTH

*Dietary Guidelines for Americans**

What should you eat to stay healthy?

Hardly a day goes by without someone trying to answer that question. Newspapers, magazines, books, radio, and television give us a lot of advice about what foods we should or should not eat. Unfortunately, much of this advice is confusing.

Some of this confusion exists because we don't know enough about nutrition to identify an "ideal diet" for each individual. People differ — and their food needs vary depending on age, sex, body size, physical activity, and other conditions such as pregnancy or illness.

In those chronic conditions where diet may be important — heart attacks, high blood pressure, strokes, dental caries, diabetes, and some forms of cancer — the roles of specific nutrients have not been defined.

Research does seek to find more precise nutritional requirements and to show better the connections between diet and certain chronic diseases.

But today, what advice should you follow in choosing and preparing the best foods for you and your family?

The guidelines below are suggested for most Americans. They do not apply to people who need special diets because of diseases or conditions that interfere with normal nutrition. These people may require special instruction from trained dietitians, in consultation with their own physicians.

*From *Nutrition and Your Health: Dietary Guidelines for Americans*, compiled by the U.S. Department of Agriculture and the U.S. Department of Health, Education and Welfare (February 1980).

These guidelines are intended for people who are already healthy. No guidelines can guarantee health or well-being. Health depends on many things, including heredity, lifestyle, personality traits, mental health and attitudes, and environment, in addition to diet.

Food alone cannot make you healthy. But good eating habits based on moderation and variety can help keep you healthy and even improve your health.

DIETARY GUIDELINES FOR AMERICANS

1. **Eat a variety of foods.**
2. **Maintain ideal weight.**
3. **Avoid too much fat, saturated fat, and cholesterol.**
4. **Eat foods with adequate starch and fiber.**
5. **Avoid too much sugar.**
6. **Avoid too much sodium.**
7. **If you drink alcohol, do so in moderation.**

□ 1. EAT A VARIETY OF FOODS □

You need about 40 different nutrients to stay healthy. These include vitamins and minerals, as well as amino acids (from proteins), essential fatty acids (from vegetable oils and animal fats), and sources of energy (calories from carbohydrates, proteins, and fats). These nutrients are in the foods you normally eat.

Most foods contain more than one nutrient. Milk, for example, provides proteins, fats, sugars, riboflavin and other B-vitamins, vitamin A, calcium, and phosphorus — among other nutrients.

No single food item supplies all the essential nutrients in the amounts that you need. Milk, for instance, contains very little iron or vitamin C. You should, therefore, eat a variety of foods to assure an adequate diet.

The greater the variety, the less likely you are to develop either a deficiency or an excess of any single nutrient. Variety also reduces your likelihood of being exposed to excessive amounts of contaminants in any single food item.

One way to assure variety and, with it, a well-balanced diet is to select foods each day from each of several major groups: for example, fruits and vegetables; cereals, breads, and grains; meats, poultry, eggs, and fish; dry peas and beans, such as soybeans, kidney beans, lima beans, and black-eyed peas, which are good vegetable sources of protein; and milk, cheese, and yogurt.

Fruits and vegetables are excellent sources of vitamins, especially vitamins C and A. Whole grain and enriched breads, cereals, and grain products provide B-vitamins, iron, and energy. Meats supply protein, fat, iron and other minerals, as well as several vitamins, including thiamine and vitamin B_{12}. Dairy products are major sources of calcium and other nutrients.

TO ASSURE YOURSELF AN ADEQUATE DIET

Eat a variety of foods daily, including selections of:

- Fruits
- Vegetables
- Whole grain and enriched breads, cereals, and grain products
- Milk, cheese, and yogurt
- Meats, poultry, fish, eggs
- Legumes (dry peas and beans)

There are no known advantages to consuming excess amounts of any nutrient. You will rarely need to take vitamin or mineral supplements if you eat a wide variety of foods. There are a few important exceptions to this general statement:

Women in their childbearing years may need to take iron supplements to replace the iron they lose with menstrual bleeding. Women who are no longer menstruating should not take iron supplements routinely.

Women who are pregnant or who are breastfeeding need more of many nutrients, especially iron, folic acid, vitamin A, calcium, and sources of energy (calories from carbohydrates, proteins, and fats). Detailed advice should come from their physicians or from dietitians.

Elderly or very inactive people may eat relatively little food. Thus, they should pay special attention to avoiding foods that are high in calories but low in other essential nutrients — for example, fat, oils, alcohol, and sugars.

Infants also have special nutritional needs. Healthy full-term infants should be breastfed unless there are special problems. The nutrients in human breast milk tend to be digested and absorbed more easily than those in cow's milk. In addition, breast milk may serve to transfer immunity to some diseases from the mother to the infant.

Normally, most babies do not need solid foods until they are 3 to 6 months old. At that time, other foods can be introduced gradually.

Prolonged breast or bottlefeeding—without solid foods or supplemental iron—can result in iron deficiency.

You should not add salt or sugar to the baby's foods. Infants do not need these "encouragements"—if they are really hungry. The foods themselves contain enough salt and sugar; extra is not necessary.

TO ASSURE YOUR BABY AN ADEQUATE DIET

Breastfeed unless there are special problems.
Delay other foods until baby is 3 to 6 months old.
Do not add salt or sugar to baby's food.

□ 2. MAINTAIN IDEAL WEIGHT □

If you are too fat, your chances of developing some chronic disorders are increased. Obesity is associated with high blood pressure, increased levels of blood fats (triglycerides) and cholesterol, and the most common type of diabetes. All of these, in turn, are associated with increased risks of heart attacks and strokes. Thus, you should try to maintain "ideal" weight.

But, how do you determine what the ideal weight is for you?

There is no absolute answer. The table on the following page shows "acceptable" ranges for most adults. If you have been obese since childhood, you may find it difficult to reach or to maintain your weight within the acceptable range. For most people, their weight should not be more than it was when they were young adults (20 or 25 years old).

It is not well understood why some people can eat much more than others and still maintain normal weight. However, one thing is definite: to lose weight, you must take in fewer calories than you burn. This means that you must either select foods containing fewer calories or you must increase your activity—or both.

TO IMPROVE EATING HABITS

Eat slowly.
Prepare smaller portions.
Avoid "seconds."

If you need to lose weight, do so gradually. Steady loss of 1 to 2 pounds a week—until you reach your goal—is relatively safe and more likely to be maintained. Long-term success depends upon acquir-

☐ SUGGESTED BODY WEIGHTS ☐

Range of Acceptable Weight

HEIGHT *(feet-inches)*	MEN *(pounds)*	WOMEN *(pounds)*
4′10″		92–119
4′11″		94–122
5′0″		96–125
5′1″		99–128
5′2″	112–141	102–131
5′3″	115–144	105–134
5′4″	118–148	108–138
5′5″	121–152	111–142
5′6″	124–156	114–146
5′7″	128–161	118–150
5′8″	132–166	122–154
5′9″	136–170	126–158
5′10″	140–174	130–163
5′11″	144–179	134–168
6′0″	148–184	138–173
6′1″	152–189	
6′2″	156–194	
6′3″	160–199	
6′4″	164–204	

Note: Height without shoes; weight without clothes.
Source: HEW conference on obesity, 1973.

ing new and better habits of eating and exercise. That is perhaps why "crash" diets usually fail in the long run.

Do not try to lose weight too rapidly. Avoid crash diets that are severely restricted in the variety of foods they allow. Diets containing fewer than 800 calories may be hazardous. Some people have developed kidney stones, disturbing psychological changes, and other complications while following such diets. A few people have died suddenly and without warning.

TO LOSE WEIGHT

Increase physical activity.
Eat less fat and fatty foods.
Eat less sugar and sweets.
Avoid too much alcohol.

Gradual increase of everyday physical activities like walking or climbing stairs can be very helpful. The chart below gives the calories used per hour in different activities.

□ APPROXIMATE ENERGY EXPENDITURE BY A 150 POUND PERSON IN VARIOUS ACTIVITIES □

Activity	Calories per Hour
Lying down or sleeping	80
Sitting	100
Driving an automobile	120
Standing	140
Domestic work	180
Walking, 2½ mph	210
Bicycling, 5½ mph	210
Gardening	220
Golf; lawn mowing, power mower	250
Bowling	270
Walking, 3¾ mph	300
Swimming, ¼ mph	300
Square dancing, volleyball; roller skating	350
Wood chopping or sawing	400
Tennis	420
Skiing, 10 mph	600
Squash and handball	600
Bicycling, 13 mph	660
Running, 10 mph	900

Source: Based on material prepared by Robert E. Johnson, M.D., Ph.D., and colleagues, University of Illinois.

A pound of body fat contains 3500 calories. To lose 1 pound of fat, you will need to burn 3500 calories more than you consume. If you burn 500 calories more a day than you consume, you will lose 1 pound of fat a week. Thus, if you normally burn 1700 calories a day, you can theoretically expect to lose a pound of fat each week if you adhere to a 1200-calorie-per-day diet.

Do not attempt to reduce your weight below the acceptable range. Severe weight loss may be associated with nutrient deficiencies, menstrual irregularities, infertility, hair loss, skin changes, cold intolerance, severe constipation, psychiatric disturbances, and other complications.

If you lose weight suddenly or for unknown reasons, see a physician. Unexplained weight loss may be an early clue to an unsuspected underlying disorder.

□ 3. AVOID TOO MUCH FAT, SATURATED FAT, AND CHOLESTEROL □

If you have a high blood cholesterol level, you have a greater chance of having a heart attack. Other factors can also increase your risk of heart attack — high blood pressure and cigarette smoking, for example — but high blood cholesterol is clearly a major dietary risk indicator.

Populations like ours with diets high in saturated fats and cholesterol tend to have high blood cholesterol levels. Individuals within these populations usually have greater risks of having heart attacks than people eating low-fat, low-cholesterol diets.

Eating extra saturated fat and cholesterol will increase blood cholesterol levels in most people. However, there are wide variations among people — related to heredity and the way each person's body uses cholesterol.

Some people can consume diets high in saturated fats and cholesterol and still keep normal blood cholesterol levels. Other people, unfortunately, have high blood cholesterol levels even if they eat low-fat, low-cholesterol diets.

There is controversy about what recommendations are appropriate for healthy Americans. But for the U.S. population *as a whole,* reduction in our current intake of total fat, saturated fat, and cholesterol is sensible. This suggestion is especially appropriate for people who have high blood pressure or who smoke.

The recommendations are not meant to prohibit the use of any specific food item or to prevent you from eating a variety of foods. For example, eggs and organ meats (such as liver) contain cholesterol, but they also contain many essential vitamins and minerals, as well as protein. Such items can be eaten in moderation, as long as your overall cholesterol intake is not excessive. If you prefer whole milk to skim milk, you can reduce your intake of fats from foods other than milk.

TO AVOID TOO MUCH FAT, SATURATED FAT, AND CHOLESTEROL

Choose lean meat, fish, poultry, dry beans and peas as your protein sources.
Moderate your use of eggs and organ meats (such as liver).
Limit your intake of butter, cream, hydrogenated margarines, shortenings and coconut oil, and foods made from such products.
Trim excess fat off meats.
Broil, bake, or boil rather than fry.
Read labels carefully to determine both amount and types of fat contained in foods.

□ 4. EAT FOODS WITH ADEQUATE STARCH AND FIBER □

The major sources of energy in the average U.S. diet are carbohydrates and fats. (Proteins and alcohol also supply energy, but to a lesser extent.) If you limit your fat intake, you should increase your calories from carbohydrates to supply your body's energy needs.

In trying to reduce your weight to "ideal" levels, carbohydrates have an advantage over fats: carbohydrates contain less than half the number of calories per ounce than fats.

Complex carbohydrate foods are better than *simple* carbohydrates in this regard. Simple carbohydrates — such as sugars — provide calories but little else in the way of nutrients. Complex carbohydrate foods — such as beans, peas, nuts, seeds, fruits and vegetables, and whole grain breads, cereals, and products — contain many essential nutrients in addition to calories.

Increasing your consumption of certain complex carbohydrates can also help increase dietary fiber. The average American diet is relatively low in fiber. Eating more foods high in fiber tends to reduce the symptoms of chronic constipation, diverticulosis, and some types of "irritable bowel." There is also concern that low fiber diets might increase the risk of developing cancer of the colon, but whether this is true is not yet known.

To make sure you get enough fiber in your diet, you should eat fruits and vegetables, whole grain breads and cereals. There is no reason to add fiber to foods that do not already contain it.

TO EAT MORE COMPLEX CARBOHYDRATES DAILY

Substitute starches for fats and sugars.
Select foods which are good sources of fiber and starch, such as whole grain breads and cereals, fruits and vegetables, beans, peas, and nuts.

□ 5. AVOID TOO MUCH SUGAR □

The major health hazard from eating too much sugar is tooth decay (dental caries). The risk of caries is not simply a matter of how much sugar you eat. The risk increases the more frequently you eat sugar and sweets, especially if you eat between meals, and if you eat foods that stick to the teeth. For example, frequent snacks of sticky candy, or dates, or daylong use of soft drinks may be more harmful than adding sugar to your morning cup of coffee — at least as far as your teeth are concerned.

Obviously, there is more to healthy teeth than avoiding sugars. Careful dental hygiene and exposure to adequate amounts of fluoride in the water are especially important.

Contrary to widespread opinion, too much sugar in your diet does not seem to cause diabetes. The most common type of diabetes is seen in obese adults, and avoiding sugar, without correcting the overweight, will not solve the problem. There is also no convincing evidence that sugar causes heart attacks or blood vessel diseases.

Estimates indicate that Americans use on the average more than 130 pounds of sugars and sweeteners a year. This means the risk of tooth decay is increased not only by the sugar in the sugar bowl but by the sugars and syrups in jams, jellies, candies, cookies, soft drinks, cakes, and pies, as well as sugars found in products such as breakfast cereals, catsup, flavored milks, and ice cream. Frequently, the ingredient label will provide a clue to the amount of sugars in a product.

TO AVOID EXCESSIVE SUGARS

Use less of all sugars, including white sugar, brown sugar, raw sugar, honey, and syrups.
Eat less of foods containing these sugars, such as candy, soft drinks, ice cream, cakes, cookies.

Select fresh fruits or fruits canned without sugar or light syrup rather than heavy syrup.

Read food labels for clues on sugar content — if the names sucrose, glucose, maltose, dextrose, lactose, fructose, or syrups appear first, then there is a large amount of sugar.

Remember, how often you eat sugar is as important as how much sugar you eat.

□ 6. AVOID TOO MUCH SODIUM □

Table salt contains sodium and chloride — both are essential elements.

Sodium is also present in many beverages and foods that we eat, especially in certain processed foods, condiments, sauces, pickled foods, salty snacks, and sandwich meats. Baking soda, baking powder, monosodium glutamate (MSG), soft drinks, and even many medications (many antacids, for instance) contain sodium.

It is not surprising that adults in the United States take in much more sodium than they need.

The major hazard of excessive sodium is for persons who have high blood pressure. Not everyone is equally susceptible. In the United States, approximately 17 percent of adults have high blood pressure. Sodium intake is but one of the factors known to affect blood pressure. Obesity, in particular, seems to play a major role.

In populations with low-sodium intakes, high blood pressure is rare. In contrast, in populations with high-sodium intakes, high blood pressure is common. If people with high blood pressure severely restrict their sodium intakes, their blood pressures will *usually* fall — although not always to normal levels.

At present, there is no good way to predict who will develop high blood pressure, though certain groups, such as blacks, have a higher incidence. Low-sodium diets might help some of these people avoid high blood pressure if they could be identified before they develop the condition.

Since most Americans eat more sodium than is needed, consider reducing your sodium intake. Use less table salt. Eat sparingly those foods to which large amounts of sodium have been added. Remember that up to half of sodium intake may be "hidden," either as part of the naturally occurring food or, more often, as part of a preservative or flavoring agent that has been added.

TO AVOID TOO MUCH SODIUM

Learn to enjoy the unsalted flavors of foods.
Cook with only small amounts of added salt.
Add little or no salt to food at the table.
Limit your intake of salty foods, such as potato chips, pretzels, salted nuts and popcorn, condiments (soy sauce, steak sauce, garlic salt), cheese, pickled foods, cured meats.
Read food labels carefully to determine the amounts of sodium in processed foods and snack items.

□ 7. IF YOU DRINK ALCOHOL, DO SO IN MODERATION □

Alcoholic beverages tend to be high in calories and low in other nutrients. Even moderate drinkers may need to drink less if they wish to achieve ideal weight.

On the other hand, heavy drinkers may lose their appetites for foods containing essential nutrients. Vitamin and mineral deficiencies occur commonly in heavy drinkers — in part, because of poor intake, but also because alcohol alters the absorption and use of some essential nutrients.

Sustained or excessive alcohol consumption by pregnant women has caused birth defects. Pregnant women should limit alcohol intake to 2 ounces or less on any single day.

Heavy drinking may also cause a variety of serious conditions, such as cirrhosis of the liver and some neurological disorders. Cancer of the throat and neck is much more common in people who drink and smoke than in people who don't.

One or two drinks daily appear to cause no harm in adults. If you drink you should do so in moderation.

For further reading on diet and its relationship to good health, send for:

Food — A publication on food and nutrition by the U.S. Department of Agriculture. Home and Garden Bulletin No. 228. Science and Education Administration. Stock No. 001–000–03881–8.
Healthy People — The Surgeon General's Report on Health Promotion and Disease Prevention. Public Health Service, U.S. Department of Health, Education, and Welfare. Stock No. 017–001–00416–2.

To obtain copies of the above publications, write to: Superintendent of Documents, U.S. Government Printing Office, Washington, D.C. 20402.

For assistance with your food and nutrition questions, contact the dietitian, home economist, or nutritionist in the following groups:

Public Health Department
County Extension office
State or local medical society
Hospital outpatient clinic
Local Dietetic Association office
Local Heart Association office
Local Diabetes Association office

Appendix 2

FOOD LABELING

*Tentative Position of Agencies**

SUMMARY

Request for Economic Information

Before taking specific action to change the current legal requirements for food labeling, the agencies want to allow further public comment on these tentative proposals. Although in some instances making these changes may require additional explicit legislation, FDA and USDA will act promptly, after fully reviewing the comments, to do what they can under their current legal authority and mandate to implement their proposals.

Detailed analysis of the economic impact of these initiatives is warranted before the agencies select specific proposals and courses of action. The agencies are therefore asking the public for information, data, or analyses to help them predict and quantify the economic impact of their proposals on industry and consumers. This information may relate to product categories, package types, characteristics of firms or establishments, manufacturing and distribution practices, etc. The agencies are particularly interested in the cost implications of the following initiatives:

Ingredient Labeling

1. Expanded label identification of ingredients in standardized foods.

2. Expanded quantitative label identification of valuable and characterizing ingredients of foods.

*From *Federal Register 44*, no. 247, pp. 76014–76017.

3. Restricted use of "and/or" labeling for fats and oils.

4. Compilation and distribution of a standard ingredient handbook or dictionary explaining the functions of ingredients used in food products.

Nutrition Labeling

1. Mandatory inclusion of sodium and potassium content as part of nutrition labeling.

2. Mandatory inclusion of sugars content as part of nutrition labeling.

3. Mandatory fatty acid labeling when cholesterol content is declared, and mandatory cholesterol declaration when fatty acid content is declared.

4. Other revisions in the mandatory list of nutrients.

5. Mandatory nutrition labeling, regardless of a food's specific content.

6. Optional use of composite data bases for nutrition labeling.

Issue	Legislation	New or Revised Regulations	Consumer Education/ Industry Guidance	Action Deferred Pending Research, Study or Future Reexamination
1. Ingredient Labeling				
A. Labeling of Mandatory ingredients in Standardized Foods.	FDA proposes to seek or support legislation to amend the FD&C Act to require declaration of mandatory ingredients in standardized foods. (The regulations issued under statutes administered by USDA already so provide.)	FDA intends to expedite amendment of the remaining food standards of identity to require declaration of all optional ingredients including the form of the mandatory ingredients where more than one form is available for use.	....	
B. Label Declaration of Colors, Spices, and Flavors.	FDA proposes to seek or support legislation to require that all colors and spices be declared on food labels by their specific name.	....	....	
	USDA believes it has authority to require the declaration of spices and colors by their specific name but will support legislation to provide more explicit authority for both agencies.	....	....	

Issue	Legislation	New or Revised Regulations	Consumer Education/ Industry Guidance	Action Deferred Pending Research, Study or Future Reexamination
B. Label Declaration of Colors, Spices, and Flavors. (*cont.*)	FDA proposes to seek or support legislation to eliminate the exemption under which the labels of butter, cheese, and ice cream are not required to declare the presence of an artificial color.	....	....	
	FDA proposes to seek or support legislation to provide it with discretionary authority to require label declaration of a flavor when it is considered necessary for providing important health information, e.g., it has a potential for causing an allergic reaction.	....	....	
	USDA believes it has authority to require declaration of flavors, but will support legislation to provide more explicit authority for both agencies.	....	....	

C. Quantitative Ingredient Labeling.	FDA proposes to seek or support legislation explicitly giving it authority to require quantitative ingredient labeling as part of the ingredient statement and to seek or support legislation giving FDA records and reports authority—access to a company's formulae, quality control records and related records in order to ensure that such expanded quantitative ingredient labeling is truthful and accurate.	FDA intends to expand the use of percentage of valuable and characterizing ingredients according to its regulations on common or usual names for nonstandardized foods.	FDA intends to publish guidelines for voluntary quantitative ingredient labeling as part of the ingredient statement.
	USDA currently has limited authority to require quantitative ingredient labeling in some cases, and it shares FDA's belief that legislation establishing the authority of the two agencies generally to require such labeling is desirable.	USDA intends to issue regulations that would provide for more quantitative ingredient labeling.	....

Issue	Legislation	New or Revised Regulations	Consumer Education/ Industry Guidance	Action Deferred Pending Research, Study or Future Reexamination
D. Ingredient Order of Predominance Statement.	....	FDA and USDA intend to amend the regulations to require that food labels bear a statement to the effect that the ingredients are listed in descending order of predominance.	....	
E. Use of "and/or" Labeling to Declare the Source of Fats and Oils.	....	FDA and USDA intend to amend the regulations to require that food containing 10 percent or more total fat on a dry weight basis declare the specific source of the fat or oil. Food containing less than 10 percent fat on a dry weight basis may use the "and/or" approach, e.g., vegetable oil (may contain cottonseed oil, soybean oil, and/or palm oil)".	....	
F. Functionality and Names of Ingredients.	....	....	....	The agencies intend to explore the concept of an "ingredient dictionary," a reference

				volume that would describe ingredient functions, and how such a publication should be compiled, distributed and funded.
G. Labeling of Fresh Fruits and Vegetables.	....	....	FDA will continue to encourage compliance with the Federal law which requires that the presence of colors, preservatives, and waxing be declared in labeling (e.g., placards and leaflets) at the point of sale of fresh fruits and vegetables.	The labeling of fresh fruits and vegetables as to pesticides and fertilizers will be reexamined in the future as necessary.
H. Ingredient Labeling for Restaurant-Served Foods.	....	....	....	The agencies will not initiate any action at this time to require that ingredients in restaurant-served foods be declared. This issue will be reexamined in the future as necessary.
I. Incidental/Secondary Ingredients.	....	....	....	The current policy of not requiring the declaration of incidental/secondary ingredients will be continued. This issue will be reexamined in the future as necessary.

Issue	Legislation	New or Revised Regulations	Consumer Education/ Industry Guidance	Action Deferred Pending Research, Study or Future Reexamination
II. Nutrition Labeling				
A. Mandatory, Discretionary, or Voluntary Nutrition Labeling.	FDA and USDA propose to seek or support legislation to clarify their authority to require nutrition labeling on all foods. (The agencies are establishing a task group to develop criteria for determining which foods should bear nutrition labeling.)	USDA intends to propose regulations that would require nutrition labeling where claims are made for a product or where certain nutrients have been added to the product.	....	
B. Nutrition Labeling Format.	....	....	The agencies encourage industry to experiment voluntarily, under controlled conditions and in collaboration with FDA and USDA, with graphics and other formats that are consistent with the current quantitative system.	FDA will conduct research, with support from USDA, to determine which format is most useful and convenient for consumers: the present nutrition labeling system will be retained pending research results and conclusions about what (if any) changes in labeling format are appropriate. (An inter-agency task group will be

				established to coordinate the research efforts with industry experimentation.)
C. Mandatory Information for Nutrition Labeling.	....	USDA intends to propose regulations that would require nutrition labeling providing information on calories, protein, carbohydrates, fat, sugars, cholesterol, sodium, and other nutrients of public health concern insofar as permitted by current authority.	....	FDA and USDA will maintain their current policies on declaring nutrients. Comments are requested on aspects of the present policies.
D. Composite Data Base for Use in Nutrition Labeling.	....	....	FDA and USDA are issuing a statement of current policy concerning use of suitable nutrient data bases in nutrition labeling.	The agencies propose to maintain the current policy that products be labeled according to composition and that it is the manufacturer's responsibility to assure the validity of nutrient content expressed on food labels. The agencies encourage industry to develop suitable data bases for use in nutrition labeling.
E. Serving Sizes.	....	FDA intends to publish final regulations for some beverage products, cereals, and meal replacements.	....	

Issue	Legislation	New or Revised Regulations	Consumer Education/ Industry Guidance	Action Deferred Pending Research, Study or Future Reexamination
E. Serving Sizes. *(cont.)*		FDA and USDA intend to propose regulations to establish serving sizes for additional product classes and/or types of foods.	....	
F. Sugars Labeling	FDA and USDA propose to seek or support legislation to provide them with explicit, discretionary authority to require quantitative labeling of sugars on the basis of public health significance.	FDA intends to amend the nutrition labeling regulations to require quantitative declaration of total sugars as part of nutrition labeling when contained above a specified level.	The agencies propose conducting an educational program to promote better understanding of declaration of sugars.	
		USDA intends to propose regulation to require the declaration of sugars content as part of nutrition labeling. (The agencies have established a task group to develop criteria for determining the threshold level for this declaration).	....	
G. Sodium and Potassium Labeling.	FDA proposes to seek or support legislation to	FDA intends to amend the nutrition labeling re-	....	

	provide it with explicit discretionary authority to require quantitative labeling of sodium and potassium on the basis of public health significance.	gulations to require sodium and potassium labeling.	
	USDA believes it has such authority, but will support legislation to provide more explicit authority for both agencies.	USDA intends to propose regulations to require sodium labeling as part of nutrition labeling. USDA will consider the inclusion of potassium labeling.	....
		FDA and USDA intend to propose regulations to define "low sodium." Consideration will be given to defining "reduced sodium" foods and standardizing the claims appropriate for food containing no added sodium.	....
H. Fatty Acid and Cholesterol Labeling.	FDA proposes to seek or support legislation to provide it with explicit discretionary authority to require cholesterol and fatty acid labeling on the basis of significance to public health.	FDA will continue to require cholesterol or fatty acid content to be included on nutrition labeling when cholesterol or fatty acid claims are made.	....

Issue	Legislation	New or Revised Regulations	Consumer Education/ Industry Guidance	Action Deferred Pending Research, Study or Future Reexamination
H. Fatty Acid and Cholesterol Labeling. (*cont.*)	USDA believes it has such authority but will support legislation to provide more explicit authority for both agencies.	FDA intends to amend the fatty acid/cholesterol regulation to require fatty acid labeling whenever cholesterol is declared and vice versa.	....	
		FDA intends to amend the fatty-acid/cholesterol regulation to eliminate the physician advice statement.	....	
		FDA intends to propose regulations to define the terms "low cholesterol," "reduced cholesterol," and "cholesterol free." FDA will consider proposing regulations to govern claims about fatty acid content.	....	
		USDA intends to propose regulations to require cholesterol labeling as part of nutrition labeling.	....	

I. Fiber Labeling	....	....	....	FDA and USDA will not require dietary fiber labeling as part of nutrition labeling until dietary fiber can be better defined, methods of analyses can be developed, and its significance in the diet determined.
J. Disease-Related Claims on Food Labels.	....	FDA intends to propose regulations covering "medical foods," that is, foods which are intended for use under medical supervision, the regulations will permit the use of appropriate medical claims about such foods.	....	FDA and USDA will maintain the present policy of not allowing disease-related claims to appear on the labeling of conventional food products. This policy will be reexamined in the future if found necessary.
III. Open Date Labeling	FDA proposes to seek or support legislation to provide it with explicit discretionary authority to require open dating on classes of foods as is found necessary.	USDA intends to propose regulations that would require open dating on perishable and semi-perishable foods to the extent that authority permits.	....	
	USDA believes it now has such authority, but will support legislation to provide more explicit authority for both agencies.	....	....	

Issue	Legislation	New or Revised Regulations	Consumer Education/ Industry Guidance	Action Deferred Pending Research, Study or Future Reexamination
IV. Imitation/Substitute Food Labeling.	....	....	....	The agencies have not taken a tentative position pending consideration of further comments on the policy direction they should follow.
V. Food Fortification	FDA and USDA propose to seek or support legislation aimed at providing explicit discretionary authority to control fortification of food when deemed of public health significance.	....	FDA will publish "General Principles for the Addition of Nutrients to Foods," and encourage manufacturers to adhere to these guidelines.	
VI. Sale and Suitable Ingredients and Food Standards.	....	FDA intends to revise its "safe and suitable" ingredients policy as follows in order to clarify the policy and ensure that it is only applied where appropriate: this policy is applicable only to optional, functional ingredients used in	....	

		quantities of less than 5 percent; the descriptive phrase "safe and suitable" will be changed to one that is less confusing and more accurately describes the concept, e.g., "permited functional ingredient."		
VII. Miscellaneous				
A. Label Declaration of Country of Origin for Imported Foods.	....	....	....	USDA and FDA propose to continue the present policy that the country of origin be declared on the label only for those foods imported and sold in consumer packages as such. This policy will be reexamined in the future as necessary.
B. Label Declaration of Name and Address of Manufacturer, Packer, and/or Distributor.	....	....	....	USDA and FDA propose to continue the present policy of requiring the declaration of the name and address of the manufacturer, packer, or distributor. This policy will be reexamined in the future as found necessary.

Issue	Legislation	New or Revised Regulations	Consumer Education/ Industry Guidance	Action Deferred Pending Research, Study or Future Reexamination
C. Natural and Organic Labeling Claims.	....	....	....	USDA and FDA propose to continue their respective policies for regulating "natural" and "organic," claims pending their evaluation of FTC's rulemaking efforts of advertising making such claims.
D. Quantity of Contents Declaration.	....	FDA will publish a final regulation.	....	

Appendix 3

NUTRITIONAL QUALITY OF FOODS

*Fortification Policy**

☐ STATEMENT OF PURPOSE ☐

(a) The fundamental objective of this subpart is to establish a uniform set of principles that will serve as a model for the rational addition of nutrients to foods. The achievement and maintenance of a desirable level of nutritional quality in the nation's food supply is an important public health objective. The addition of nutrients to specific foods can be an effective way of maintaining and improving the overall nutritional quality of the food supply. However, random fortification of foods could result in over- or underfortification in consumer diets and create nutrient imbalances in the food supply. It could also result in deceptive or misleading claims for certain foods. The Food and Drug Administration does not encourage indiscriminate addition of nutrients to foods, nor does it consider it appropriate to fortify fresh produce; meat, poultry, or fish products; sugars; or snack foods such as candies and carbonated beverages. To preserve a balance of nutrients in the diet, manufacturers who elect to fortify foods are urged to utilize these principles when adding nutrients to food. It is reasonable to anticipate that the U.S. RDA's as delineated in § 101.9 of this chapter and in paragraph (d) of this section will be amended from time to time to list additional nutrients and/or to change the levels of specific U.S. RDA's as improved knowledge about human nutrient

*From *Federal Register 45*, no. 18, pp. 6323–6324.

requirements and allowances develops. The policy set forth in this section is based on U.S. dietary practices and nutritional needs and may not be applicable in other countries.

(b) A nutrient(s) listed in paragraph (d)(3) of this section may appropriately be added to a food to correct a dietary insufficiency recognized by the scientific community to exist and known to result in nutritional deficiency disease if:

(1) Sufficient information is available to identify the nutritional problem and the affected population groups, and the food is suitable to act as a vehicle for the added nutrients. Manufacturers contemplating using this principle are urged to contact the Food and Drug Administration before implementing a fortification plan based on this principle.

(2) The food is not the subject of any other Federal regulation for a food or class of food that requires, permits, or prohibits nutrient additions. (Other Federal regulations include, but are not limited to, standards of identity promulgated under section 401 of the Federal Food, Drug, and Cosmetic Act, nutritional quality guidelines established in Subpart C of this part, and common or usual name regulations established in Part 102 of this chapter.)

(c) A nutrient(s) listed in paragraph (d)(3) of this section may appropriately be added to a food to restore such nutrient(s) to a level(s) representative of the food prior to storage, handling, and processing, when:

(1) The nutrient(s) is shown by adequate scientific documentation to have been lost in storage, handling, or processing in a measurable amount equal to at least 2 percent of the U.S. RDA (and 2 percent of 2.5 grams of potassium and 4.0 milligrams of manganese, when appropriate) in a normal serving of the food;

(2) Good manufacturing practices and normal storage and handling procedures cannot prevent the loss of such nutrient(s).

(3) All nutrients, including protein, iodine and vitamin D, that are lost in a measurable amount are restored and all ingredients of the food product that contribute nutrients are considered in determining restoration levels; and

(4) The food is not the subject of any other Federal regulation that requires or prohibits nutrient addition(s), or the food has not been fortified in accordance with any other Federal regulation that permits voluntary nutrient additions.

(d) A nutrient(s) listed in paragraph (d)(3) of this section may be added to a food in proportion to the total caloric content of the food, to balance the vitamin, mineral, and protein content if:

(1) A normal serving of the food contains at least 40 kilocalories (that is, 2 percent of a daily intake of 2,000 kilocalories);

(2) The food is not the subject of any other Federal regulation for a food or class of food that requires, permits, or prohibits nutrient additions; and

(3) The food contains all of the following nutrients per 100 kilocalories based on a 2,000-kilocalorie total intake as a daily standard:

NUTRIENT AND UNIT OF MEASUREMENT	U.S. RDA[1]	AMOUNT PER 100 KILOCALORIES
Protein (optional), gram (g)	65[2]	3.25
	45	2.25
Vitamin A, international unit (IU)	5000	250
Vitamin C, milligram (mg)	60	3
Thiamine, milligram (mg)	1.5	0.075
Riboflavin, milligram (mg)	1.7	0.085
Niacin, milligram (mg)	20	1.0
Calcium, gram (g)	1	0.05
Iron, milligram (mg)	18	0.9
Vitamin D (optional), international unit (IU)	400	20
Vitamin E, international unit (IU)	30	1.5
Vitamin B–6, milligram (mg)	2	0.1
Folic acid, milligram (mg)	0.4	0.02
Vitamin B–12, microgram (mcg)	6	0.3
Phosphorus, gram (g)	1	0.05
Iodine (optional), microgram (mcg)	150	7.5
Magnesium, milligram (mg)	400	20
Zinc, milligram (mg)	15	0.75
Copper, milligram (mg)	2	0.1
Biotin, milligram (mg)	0.3	0.015
Pantothenic acid, milligram (mg)	10	0.5
Potassium, gram (g)	(3)	0.125
Manganese, milligram (mg)	(3)	0.2

[1] U.S. Recommended Daily Allowance (U.S. RDA) for adults and children 4 or more years of age.
[2] If the protein efficiency ratio of protein is equal to or better than that of casein, the U.S. RDA is 45 g.
[3] No U.S. RDA has been established for either potassium or manganese; daily dietary intakes of 2.5 g. and 4.0 mg., respectively, are based on the 1979 Recommended Dietary Allowances of the Food and Nutrition Board, National Academy of Sciences-National Research Council.

(e) A nutrient(s) may appropriately be added to a food that replaces traditional food in the diet to avoid nutritional inferiority in accordance with § 101.3(e)(2) of this chapter.

(f) Nutrient(s) may be added to foods as permitted or required by applicable regulations established elsewhere in this chapter.

(g) A nutrient added to a food is appropriate only when the nutrient:

(1) Is stable in the food under customary conditions of storage, distribution, and use;

(2) Is physiologically available from the food;

(3) Is present at a level at which there is a reasonable assurance that consumption of the food containing the added nutrient will not result in an excessive intake of the nutrient, considering cumulative amounts from other sources in the diet; and

(4) Is suitable for its intended purpose and is in compliance with applicable provisions of the act and regulations governing the safety of substances in food.

(h) Any claims or statements in the labeling of food about the addition of a vitamin, mineral, or protein to a food shall be made only if the claim or statement is not false or misleading and otherwise complies with the act and any applicable regulations. The following label claims are acceptable:

(1) The labeling claim "fully restored with vitamins and minerals" or "fully restored with vitamins and minerals to the level of unprocessed — " (the blank to be filled in with the common or usual name of the food) may be used to describe foods fortified in accordance with the principles established in paragraph (c) of the section.

(2) The labeling claim, "vitamins and minerals (and "protein" when appropriate) added are in proportion to caloric content" may be used to describe food fortified in accordance with the principles established in paragraph (d) of this section.

(3) When labeling claims are permitted, the term "enriched," "fortified," "added," or similar terms may be used interchangeably to indicate the addition of one or more vitamins or minerals or protein to a food, unless an applicable Federal regulation requires the use of specific words or statements.

(i) It is inappropriate to make any claim or statement on a label or in labeling, other than in a listing of the nutrient ingredients as part of the ingredient statement, that any vitamin, mineral, or protein has been added to a food to which nutrients have been added pursuant to paragraph (e) of this section.

Effective date. This policy statement is effective February 25, 1980. (Secs. 201(n), 403(a), 701(a), 52 Stat. 1041 and 1047 as amended, 1055

(21 U.S.C. 321(n), 343(a), 371(a)).); Dated: January 18, 1980. Jere E. Goyan, Commissioner of Food and Drugs. [FR Doc. 80–2380 Filed 1-24-80; 8:45 am]; Billing Code 4110–03–M.

Appendix 4

UNDERSTANDING NUTRITION LABELS*

Naturally you want to get the best possible nutrition from the foods you buy. But you can't tell how nutritious a food is just by looking at it. That's where nutrition labeling comes in.

Under FDA regulations any food to which a nutrient has been added, or any food that makes a nutritional claim, must have a nutrition label. And many food manufacturers put nutrition labels on other foods.

If you understand how to read these labels, they can help you get better nutrition and save money, too. Using nutrition labels, you can:

Choose enough of the nutrients you need daily. If you want to be sure you're getting enough vitamin A, for example, check food labels to find which foods are good sources of this vitamin.

Count calories. All nutrition labels tell how many calories are in a serving of the food.

Avoid foods containing substances such as cholesterol or sodium which your physician may have advised you to cut down on.

Compare different brands of the same product to find the one that gives you the most nutrition for your money. If different brands contain about the same nutrients, the least expensive is the best buy, nutrition-wise.

*From U.S. DEPARTMENT OF HEALTH, EDUCATION AND WELFARE, Public Health Service, Food and Drug Administration, Office of Public Affairs, Rockville, Maryland 20857. HEW Publication No. (FDA) 77–2082

Choose less costly, but equally nutritious, substitutes for more expensive foods. For instance, many canned and packaged foods such as dried peas and beans are good sources of protein at reasonable prices.

Learn what's in new foods you see on the grocery shelf, such as soy products or snack foods, and compare them with the nutritional quality of familiar foods.

This brochure tells what you'll see on nutrition labels, and what the information means. It will help you to understand nutrition labels so you can use them in buying foods and in planning meals. The nutrition labels used in this brochure are samples from actual food products.

□ SERVING INFORMATION □

Since nutrition information is given per serving, naturally you need to know how large one serving is — 1/2 cup, 2 ounces, 1 slice, whatever. So that's the first item on the nutrition label. If the serving size is listed as "1 cup" it means that 1 cup of the food is considered a reasonable serving for an adult. That amount may be more, or less, than you or some family members usually eat. You'll have to take this into account when you look at the nutrition information. If the serving size says "1 cup," but about 1/2 cup is the amount you usually eat, then the food will provide only half as many calories and nutrients as shown on the label.

Next comes the number of servings in the package or can or container. If the serving size is 2 ounces, and there are 4 servings per container, that means there are four 2-ounce servings of the product in the container.

□ WHAT'S IN A SERVING □

The next part of the Nutrition Label tells you the number of calories and the amount of protein, carbohydrate, and fat in each serving of the product. Calories are listed first. The label might say: *Calories . . . 110,* which means there are 110 calories in each serving of the food.

Protein is listed next, followed by carbohydrate and then fat, with the amount of each of these shown in grams (g). For example, if the label shows: *Protein . . . 2 g.,* that means there are two grams of protein in one serving of the product. If you want to translate grams into ounces, 28 grams equal one ounce.

□ PERCENTAGE OF U.S. RECOMMENDED DAILY ALLOWANCES (U.S. RDA'S) □

You can't read and understand this part of the label unless you know what the U.S. Recommended Daily Allowances are. They are a guide to the amounts of protein, vitamins, and minerals you need each day for good nutrition. The U.S. RDA's most commonly used on nutrition labels are for adults and children four or more years of age. They represent approximately the amounts needed by a healthy adult male. Hence, they are higher than the needs of some people, especially younger children. As a guide, the foods you eat each day should provide reasonably close to 100 percent of the U.S. RDA's.

Special U.S. RDA's are used on baby foods and other products intended for small children. These represent the needs of infants and children under four years.

Under the U.S. RDA the label shows the amounts of protein, vitamins, and minerals in foods as percentages of the U.S. RDA. So, if the label shows: *Protein . . . 20,* it means that a serving of the size shown in the upper part of the label provides 20 percent of the protein needed in a day. You also need to eat some other foods that contain protein in order to get 100 percent of the amount recommended for a day. (Note: don't confuse the percentage of protein shown in this section with the amount [grams] of protein listed in the upper part of the label.)

■ VITAMINS AND MINERALS

The next seven items listed on the label are vitamins and minerals. The nutrition label must list five vitamins — vitamin A, vitamin C, and three of the B vitamins: thiamine, riboflavin, and niacin. If the label shows: *Niacin . . . 2,* it means one serving of the product contains 2 percent of the niacin you should have in one day. Obviously you need a lot more niacin from other foods during the day to get 100 percent of the recommended daily allowance.

The two minerals that must be listed on nutrition labels are calcium and iron. If the label shows: *Iron . . . 25,* it means one serving of the product will give you 25 percent of the iron needed in a day.

Not all foods contain all of the seven vitamins and minerals that must be listed on nutrition labels. In fact, some foods contain only one or two of the vitamins or minerals that must be listed on nutrition labels. If a food contains less than 2 percent of the U.S. RDA per serving of a vitamin or mineral, the label shows a zero or an asterisk for that nutrient: *Vitamin C . . . 0,* or *Vitamin C . . .* *.

The asterisk means the food contains less than 2 percent of the nutrient per serving, and an asterisk and statement will indicate this at the bottom of the label.

On foods which do not contain all of the vitamins and minerals that must be listed, the label may simply list the ones the product contains and state that it does not contain the others. For example, the label may show: *Vitamin C . . . 40,* and then say: *Contains less than 2 percent of the U.S. RDA of vitamin A, thiamine, riboflavin, niacin, calcium, and iron.*

Several additional essential vitamins and minerals may be listed in this section, but are not required to be listed by FDA. These are vitamin D, vitamin E, vitamin B_6, folic acid, vitamin B_{12}, phosphorus, iodine, magnesium, zinc, copper, biotin, and pantothenic acid. If you get enough of the seven vitamins and minerals which must be listed on the label you usually will get enough of these additional nutrients.

□ OTHER INFORMATION THAT IS NOT REQUIRED BUT IS SOMETIMES GIVEN □

■ CHOLESTEROL AND FAT

To help people who want to alter their fat intake or cut down on cholesterol, some manufacturers list cholesterol and provide more detailed information on fat content than FDA requires. Here's how these optional items are shown:

Cholesterol

When cholesterol content is given, the amount of cholesterol in each serving, and also the amount in each 100 grams, of the product is shown: *Cholesterol (14 mg/100 g) . . . 34 mg.*

This means there are 14 milligrams of cholesterol in each 100 grams of the product and 34 milligrams in each serving of the product.

Fat

Some labels tell you what percent of the total calories in the product are supplied by fat and how much of the fat is polyunsaturated and how much is saturated. A label that gives this information might show:

Fat (Percent of Calories 53%)	33 G
Polyunsaturated	2 G
Saturated	9 G

This means that 53 percent of the total number of calories in one serving of the product is supplied by fat. The rest of the calories are supplied by protein or carbohydrate. One serving of the product contains 33 grams of fat. Of this, 2 grams are polyunsaturated, 9 grams are saturated, and the rest of the fat is monounsaturated. The Nutrition Label must show the total fat content, but breaking it down into polyunsaturated and saturated fat and showing the percent of calories supplied by fat is optional.

When cholesterol content is given and fat content is broken down into polyunsaturates and saturates, the label must carry the following statement: Information on fat and/or cholesterol content is provided for individuals who, on the advice of a physician, are modifying their dietary intake of fat and/or cholesterol.

■ SODIUM

Some food manufacturers put the sodium content on the label for the benefit of people who are reducing the sodium in their diets on the advice of their physicians. Commonly, it is said that salt is restricted in the diet but sodium (which is part of salt) is the substance that is actually restricted. The information on sodium is shown in milligrams per 100 grams of the product, the same way that cholesterol is shown.

INDEX